Conserving the Emerald Tiger

Conserving the Emerald Tiger:
The Politics of Environmental Regulation in Ireland

George Taylor

Arlen House
2001

© George Taylor, 2001

All rights reserved.

first published in September 2001 by

Arlen House
PO Box 222
Galway

and

42 Grange Abbey Road
Baldoyle
Dublin 13
Ireland

International Standard Book Numbers:
1–903631–13–X paperback
1–903631–14–9 hardback

www.arlenhouse.ie

Design by Dunleavy Design, Salthill, Galway
Printed by ColourBooks, Baldoyle, Dublin 13

CONTENTS

List of Abbreviations xi

Acknowledgements xii

Introduction 1

Chapter One 9
'Accomodations to Reality':
Administrative Conventions, Regulatory Practices and
Environmental Policy 'Irish Style'

Chapter Two 35
'A Watchdog, not a Bloodhound':
The Formation, Structure and Regulatory Ethos of the
Environmental Protection Agency

Chapter Three 73
'Custodian, Guardian or Absentee Landlord':
Environmental Management and the EPA in the 1990s

Chapter Four 119
'Politics Dressed up as Science':
Political Participation, Environmental Democracy and
the EPA

Bibliography 160

Notes 171

Index 176

Acknowledgements

The author would like to take this opportunity to thank Cathi Murphy, Avril Horan, Michelle Millar, Adrienne O'Reilly, Sarah Reenan, Una Burns and Ricca Edmondson for their invaluable comments at various stages of the preparation of the manuscript.

Alan Hayes, the editor, also deserves praise for his useful comments, observations and all round professionalism which made the publication of this book possible.

I would also like to record my gratitude to the Millennium Research Fund of the National University of Ireland, Galway for funding which contributed to the research behind this book.

List of Abbreviations

BAT – Best Available Technology
BATNEEC – Best Available Technology Not Exceeding Excessive Cost
BPM – Best Practicable Means
BSE – Bovine Spongiform Encalopathy
CAP – Common Agricultural Policy
CEGB – Central Electricity Generating Board
CFYPS – Control Farmyard Pollution Scheme
EIA – Environmental Impact Assessment
EIS – Environmental Impact Statement
EM – Ecological Modernisation
FICI – Federation of Irish Chemical Industries
GMO – Genetically Modified Organism
IBEC – Irish Business and Employers Confederation
IDA – Industrial Development Authority
IFA – Irish Farmers Association
IIRS – Institute For Industrial Research and Standards
IPC – Integrated Pollution Control
IPCC – Irish Peatland Conservation Council
NHA – National Heritage Areas
NIMBY – Not In My Back Yardism
nvCJD – New Variant Creuztfeld Jacob Disease
REPS – Rural Environmental Protection Scheme
SAAO – Special Amenity Area Orders
SAC – Special Area of Conservation

to

Deirdre J

with all my love.

Conserving the Emerald Tiger

Introduction

> Acid rain, BSE, dioxins, eutrophication, phosphates, **Askeaton**, **Lough Leane**, **Killarney**. Architectural heritage, bungalow 'blitz', An Bord Pleanála, **Dublin**. **Achill island**. *Lancefort*. Water pollution, sewerage, Mutton Island, *Save Galway Bay*. Genetically modified organisms, food labelling Monsanto, *Genetic Concern*. IPC licensing, Masonite, Merrell Dow, Raybestos Manhattan, Procter and Gamble**, Nenagh, Leitrim.** Landfills *Cork Environmental Alliance, the Clare Alliance Against Incineration.* Wind farms, Blue flags, interpretative centres, mobile phone masts, roads, the corncrake, REPS, SACs, IFA. **Mullaghmore**, **Connemara**, **Mayo**, the **Glens of Wicklow**. *Greenpeace, Voice*, *eco-warriors*, the Environmental Protection Agency …

As a youngster I was transfixed by the image of Henry Kissinger, complete with a grey suit, FBI glasses and a smile, waving to an 'adoring' crowd as he stepped off a 747 in some war torn zone intent on brokering peace. My vocabulary was effused with the names of Vietnam, Phnom Phenn, Saigon, napalm and agent orange. I had little understanding of these events, why they were so prominent or, more importantly, any explanation. Kissinger's presence on the Six O'Clock News was, it seems, enough to convey their importance.

Today, it appears that our attention is drawn ever more to the environment. Indeed, it is almost impossible to pick up a paper or watch the news without coming across a reference to the continuing problems of Ireland's ecology; from acid rain to genetically modified foods, Achill island to Uggool beach, our vocabulary is immersed in the issues of environmental politics. We are all familiar with the places and protagonists of this debate; from the Ecowarriors in the Glens of Wicklow to Mullaghmore and the Burren Action Group. On the more complicated issues, often draped in an impenetrable scientific language, we are perhaps more circumspect. These are difficulties not always confined to the lay public. Presiding over the recent case in which Genetic Concern challenged the Environmental Protection Agency (EPA) over genetically modified organism (GMO) trials for Monsanto, Mr Justice O'Sullivan asked both sides if they could agree to a simplified guide to genetic engineering to which he could refer *(Irish Times*, 16 July 1998).

There are undoubtedly some who are sufficiently confident in their understanding of the link between BSE and nvCJD or fish kills and eutrophication. Others, however, are prepared to rely more upon an intuitive 'feel' that the excessive use of fertilisers leads to eutrophication, even if they do not understand the scientific processes at hand. Yet others, perhaps more wary of the 'objective role' of science following the BSE crisis, have a 'gut reaction' that something is awry. Science, or at least some scientists, may confidently declare that genetically modified food is harmless, they may even go as far as to demonstrate this by eating such produce. However, the image of the former British Minister, John Gummer, feeding his children on hamburgers is still fresh enough to inform the sceptical mind.

This book suggests that any explanation of these disparate issues needs to recognise that the environment is a political realm. At first, this appears a rather strange turn of phrase for it is more common to suggest that environmental politics or the 'green issue' is about conservation or protecting endangered species and threatened habitats. And it is. But if we are to understand why the environment is a political realm we need to rid ourselves of the view that Irish politics is simply about what takes place in the Dáil. The politics of the environment embraces issues which are at the very heart of Irish democracy: the appropriate role of state intervention; the struggle to secure economic growth and conserve the environment; the anxiety of unemployment and the problems of social exclusion; the freedom of information which allows us to scrutinise and bring to account public organisations such as the EPA; public concern about genetically modified foods; the fear of dioxin release; political protest at the location of landfills. These are real issues, environmental issues which are public issues and, consequently, they are necessarily *political* issues.

In a nutshell, this book examines the changing nature of environmental politics in Ireland and provides an odyssey through Ireland's environmental regulatory framework. A cumbersome phrase, I know, but an important one. The more common perception is that a regulatory framework (or regime) refers simply to the mass of legislation which surrounds the environment; from planning laws which allow us to build or extend houses to limits on toxic waste emissions. On one level it certainly involves this, but it also contains the more complicated issue of how the legislation is implemented and used. In other words, there is a significant difference between the *intent* of legislation and the *practice* of environmental protection. For example, the EPA Act states that environmental protection includes the 'prevention, limitation, elimination, abatement or reduction of environmental pollution' (EPA Act, 1992, p. 10). Clear and comprehensive. Even reassuring. However, on closer inspection we find that the Environmental Protection Agency's operating rationale is not concerned exclusively with the *enforcement* of stringent or rigid environmental standards. Rather, the agency prefers to adopt a (soft) regulatory relationship with industry, one which avoids confrontation and tolerates variances from good environmental management as long as the 'overall environmental performance is good' (Maclean, 1994, p. 82). As one of its directors observes, while the non-enforcement of licence conditions cannot be seen as constituting good environmental management, 'actual enforcement comes in a *number of different forms*' (Maclean, 1994, p. 82, emphasis added).

The agency's preferred practice is to rely on voluntary compliance, cajole those who operate under its licences to practise 'good environmental management' and, where necessary, encourage more environmental awareness. This extends to the agricultural community where the EPA believes that the:

> challenge is to *encourage* an environmental ethos in the *minds and hearts* of farmers so that they will make the right choices for environmental protection (Sherwood, 1994, p. 66, emphasis added).

It is an approach which involves a subtle, but nonetheless crucial, shift from environmental protection to environmental management, one in which a political dynamic is at work between the agency (regulator) and those it regulates; industry, farmers, local authorities and property developers. The political will to enforce stringent (and improving) environmental standards in such a context is often compromised by resistance from those who are 'politically well organised and connected' or those upon which the community is economically dependent. Such dependence upon a large employer can often hinder even the most ardent attempts by local authorities to introduce more stringent legislation when a company believes it will impair its competitiveness. The threat, potential or otherwise, of a multi-national company seeking to relocate production presents a very real dilemma for those who need to achieve a balance between measures to ensure job creation and environmental regulation. The ability of multi-national companies to relocate production is compounded by competition among both nations and regional authorities to court investment. The environmental standards enshrined in a regulatory framework (and the determination to enforce them) emerge as simply one factor among many in the decision making of multi-national companies. There is the prospect that they will locate investment where regulation is lax, seeking out pollution havens.

Instances where licences are reviewed or contested also raise the thorny issue of environmental democracy, or the extent to which the decision making process is open to concerned citizens and pressure groups. Here, access to environmental information, the nature and extent of pollution monitoring undertaken by the EPA and the role of civil courts are all important elements which influence the public's confidence in the environmental regulatory framework. In recent years this confidence in Ireland's regulatory framework has been progressively undermined, stimulating an upsurge in environmental protest on issues as disparate as waste disposal, genetic engineering, the expansion of the road network and architectural heritage.

In this context the book argues that the EPA represents a response to a series of environmental conflicts in the late 1980s which had begun to undermine the traditional influence of the agricultural and business lobbies. Indeed, the book contends that while the Irish State had recognised the need to revamp environmental policy its principal, overarching objective was to ensure that further, more stringent regulation would not be detrimental to the economic performance of the Emerald Tiger. Put simply, environmental policy debate in Ireland is concerned no longer with the extent of ecological degradation, the quality of the environment or encouraging environmental sensitivity, but with the complicated process of organising consent around new definitions of the *extent to which pollution can be justified* (Taylor, 1998). The profusion of environmental protest groups (and the increasingly sophisticated nature of that protest) is ample testimony of the failure of the Irish environmental regulatory framework to respond to the problems which have emerged in the 1990s.

The book is divided into four chapters and largely follows the advice of a good friend who often said 'Tell me the story George, before you give me the theory'. Where possible, it also strives to avoid unnecessary jargon. The first chapter examines the regulatory framework which existed prior to the EPA and charts the emergence of an environmental crisis in the late 1980s. In particular, it focuses upon the influential role performed by the agricultural and business lobbies who were actively engaged in the reconstruction of a new environmental regulatory framework which would accommodate the environmentalist critique of the 1980s without threatening the free market ethos which has underpinned the Emerald Tiger.

The second chapter details the emergence, formation and implementation of the environmental regulatory framework which has evolved since 1992. For many, the arrival of the EPA was greeted as a new departure in Irish politics, displaying a willingness to finally accord the environment a pivotal role in public policy. It was envisaged that the agency would replace the rigid demarcation of bureaucratic responsibilities between different public bodies with a flexible, integrated and administratively transparent institution. Its responsibilities would be extensive, incorporating guidelines and backup to local authorities and other public bodies; co-ordinate environmental monitoring and publish reports on the state of the environment; undertake environmental research; supervise the environmental performance of local authorities; participate in environmental impact assessments; promote and develop environmental audits; create an eco-labelling scheme for ecologically friendly services and finally, issue and approve environmental quality objectives and codes of practice.

In contrast to the euphoria and rhetoric which accompanied the arrival of the Environmental Protection Act, this book contends that the agency has impacted significantly on the regulatory style of environmental policy in Ireland. In this context, the third chapter focuses critically upon the performance of the regulatory regime in four key areas: agriculture, industry, planning and waste management.

The fourth and final chapter explores the problems and possibilities for future policy development. It argues that there are two inextricably linked sources of tension which reside in the heart of environmental regulatory regime which need to be resolved: the need to reconcile the relationship between economic growth and further environmental protection and the imperative to establish greater political participation, transparency and accountability.[1]

8

'Accommodations to Reality': Administrative Conventions, Regulatory Practices and Environmental Policy 'Irish Style'.[2]

While the primary responsibility for the direction of environmental policy in Ireland during the 1970s and 1980s lay with the Department of the Environment, local authorities were the principal agents responsible for the implementation of legislation. The Local Government (Water Pollution) Act, 1977 combined with the Local Government (Planning and Development) Acts of 1963 and 1976 and the Local Government (Planning and Development) Regulations of 1977 and European Directives, made local authorities responsible for air and water pollution controls, sanitation and waste management. The Air Pollution Act, 1987 was a consolidation of the various EC Directives and retained local government responsibility for enforcement of air pollution controls. Most other EEC Directives relating to the environment were incorporated into the Irish environmental regime by making government regulations under the EEC Act, 1972. Although the Planning Acts were not designed specifically for the task of pollution control they were, in the absence of more effective measures, the most significant pieces of legislation (Scannell, 1982).

If policy was devolved generally to local authorities, contributing to the sense of the green issue as a local political issue, it was also developed largely on an ad hoc, reactive and incremental basis. Legislation seemed to follow a well worn path where governments simply 'added on' new elements to the remit of local authorities. For example in 1975, when the EEC decided to adopt a common procedure for the exchange of information about monitoring, the Department of the Environment issued a circular instructing local authorities to attach conditions to planning permission, a move designed to ensure that monitoring would be undertaken by developers where air pollution problems were anticipated (Scannell, 1982, p. 80).

A further feature of policy in this period was that it avoided the onerous task of either specifying precise environmental standards or defining the principles which guided controlling authorities. There have been a succession of 'principles' which have guided practice on environmental regulation. The Alkali Works Regulation Act of 1906 adopted the term 'best practicable means' when attempting to make decisions on suitable production practices. The environmental lobby have always contended that what is best (which is contentious in technical and legal terms) and what is 'practicable' will, more often than not, hinge upon how much they cost. In its most recent guise (EPA Act, 1992) this has been replaced by the Best Available Technology Not Entailing Excessive Cost (BATNEEC). Here, a subtle but nonetheless important change has taken place, in which the onus shifts to the polluter to justify the choice of technology on commercial as well as technical grounds. However, as we shall see below, a difficulty still resides in the way in which decisions seek to resolve the tension between BAT & NEEC. Whenever possible, legislation was reluctant to curtail the autonomy of regulatory bodies, thereby sustaining a working environment in which negotiated compliance remained the 'order of the day'. Within this consensual style of regulation it was not uncommon for close consultation to take place between the regulator and business interests. In essence, it was an accommodative political style which acknowledged the need to 'operate in the real world' where pragmatic, as opposed to aspirational, responses were necessary. Here, Clientilism is one of a number of terms used which attempt to unravel the processes which influence the democratic nature of a particular society. For Chubb (1970), the persistence of patronage and brokerage, which he thought were anomalies in modern Ireland, were due to a political culture that for generations had seen that:

to get the benefits that public authorities bestow, the help of a man with connections and influence was necessary. All that democracy had meant is that such a man has been laid on officially, as it were, and is now no longer a master but a servant (cited in M. D. Higgins, 1982).

In contrast, pluralism, the dominant political tradition of post-war America, argued that policy outcomes reflect the interaction between a diverse range of overlapping interest groups and government. It is this interaction or conflict which becomes the defining feature of a modern society (often termed a polyarchy) and which distinguishes itself from parliamentary forms of political representation (where your vote is the predominant influence) and single party socialist states. There are at least two important themes to this 'pluralist' position, both of which later received criticism. The first, that government remains a neutral actor, favouring no particular group, and second, that no one interest group should dominate over a long period of time. This is evident in the vocabulary of pluralists which evinces groups as opposed to classes, incremental as opposed to revolutionary change, influence rather than power, and negotiation/consultation rather than imposition. The important point here is that pluralists assign a positive role to interest groups who ensure that the liberal democratic values of political diversity and conflict are upheld.

These themes are crucial and inform our later discussion of ideas such as political access, participation and influence. In its more contemporary variant the concept of 'policy community' has been elicited. Here, once again, there is an attempt to explore the manner in which interest groups vie with each other (and government) to influence the policy process. In contrast to the earlier versions of pluralism, which emphasised diversity and neutrality (and which downplayed the power differences between groups), this approach attempts to capture the manner in which groups (of varying power) gain privileged access, secure benefits and achieve continuity in policy. Consequently, exclusion, or the absence of particular groups from the decision making process, becomes an important focus since it has a crucial bearing upon the character of subsequent policy. Environmental regulatory issues certainly proved a fertile terrain upon which clientilist politics flourished.[3] As Gilmore's exchange in the Dáil testifies, political representation along these lines could hold a profound influence on policy (and its enforcement):

> There is a problem, in that before an officer from a local authority returns to his office or lab after taking a sample from a farm or an enterprise, some Fine Gael or Fianna Fáil councillor, or perhaps TD will have made representations by telephone not to proceed with a prosecution. Local authorities and the Regional Fisheries Boards are bombarded with political representations not to pursue pollution prosecutions. That is the political culture in which we live (Gilmore, *Dáil Debates*, 1990, vol. 394, p. 637).

It was then a period in which Irish policy making conformed to a policy style not too dissimilar to that of the UK, accentuating the importance of procedural regularity, the significance of administrative convention and the value of consultation.[4] With an over-riding imperative to avoid conflict, reform, where it took place, was largely incremental and undertaken only after extensive consultation with organised interest groups.

From civil service structures to the administrative culture, there were clear similarities with the British style in terms of the way problems were addressed and ultimately resolved. It should surprise few then that environmental policy in Ireland suffered problems similar to those experienced in the UK, where policy was hampered by a lack of co-ordination, integration and a parliamentary style in which a Ministers' allegiance (and ultimately political prestige) was defined by her or his success in defending a particular Department. Although negotiations preceding the 'budget' are commonly cited to illustrate these divisions, it is often with the more mundane, 'everyday' features of parliamentary activity that its debilitating influence is revealed. This can be seen, for example, in a recent exchange in the Dáil where it emerged that it was by no means clear whether the regulatory responsibilities for tree planting lay with the Department of Agriculture, Food and Forestry (DAFF) or the Department of the Environment (DoE). When questioned in the Dáil on this matter, the Minister for the Environment was of the opinion that the decision on the type of trees planted was primarily a matter for DAFF, while environmental impact and planning requirements were in the jurisdiction of the DoE (Howlin, *Dáil Debates*, vol. 465, p. 458). However, the Minister of State at DAFF, when asked who would be responsible for environmental impact and planning controls of forestry, declared that the Forestry Department would be responsible for *all matters* effecting the forestry programme. When it was pointed out that the Minister of the Environment had seen certain aspects of the forestry programme falling within the remit of DoE, the Minister repeated that 'our Department will be responsible for *all matters that we deem affect the implementation of our programme*' (Deenihan, *Dáil Debates*, vol. 454, p. 1334; emphasis added).

A further feature endemic to the Irish policy style is the importance assigned to consultation. On the one hand, formal and informal contacts with organised interests are an important source of political legitimacy, offsetting the possibility of revolt and, on the other hand, acting as an invaluable source of information for policy judgements. In other words, such relationships are generally reciprocal and are often based upon the maxim that 'only the wearer knows where the shoe pinches'. Granting access to organised interest groups in this way simply makes the system 'more effective in supplying public needs' (Richardson and Jordan, 1982). And yet, there is also the discernible inclination in Irish politics to move beyond consultation toward negotiation.

On this subject it is clear that within the Irish political milieu not all interest groups are treated 'equally'. When considering the implementation of the EU Habitats Directive, for example, consultation with the Irish Peatland Conservation Council, (IPCC) and the Department of the Arts Culture and the Gaeltacht extended to one meeting before Christmas 1996, at which the IPCC were simply shown a draft copy of the regulations and told that they could not be changed. And yet, almost two days of Partnership 2000 talks were devoted to discussions with the agricultural lobby with regard to the issue of compensation (*Irish Times*, 3 Jan. 1997). It is one thing to be allowed to enter the office, quite another to be allowed to stay and influence decisions.

Any explanation of environmental policy therefore needs to consider the importance of political access and the manner in which it is secured by organised interest groups such as the Irish Farmers Association (IFA). There is a crucial political dynamic at work here, one in which policy is influenced (and shaped) by the government's desire either to avoid political conflict or advance its interests. This involves more than simply an occasional political favour to secure temporary political respite for a minority coalition, a feature synonymous with the current political administration where a political tirade followed the lobbying efforts of Independent Kerry TD Jackie Healy Rae to include Kerry and Clare in Ireland's EU regional funding submission (*Irish Times*, 17 Nov. 1998). It has also occurred over the location of mobile phone masts at Garda stations where Mildred Fox, TD for Wicklow, pressed for an agreement similar to that offered to Harry Blaney, independent TD for Donegal, where the Minister for Justice offered to stall the erection of telephone communications. Rather, it is about granting 'privileged' political access to organised interests who regularly contribute to an agenda formed in a tightly closed 'policy community'.

A policy community is characterised by a limited number of participants where members frequently interact and help to establish a continuity in views and reinforce consensus. The emphasis is upon how policy is formed and sustained. The community will have an influence on, although it clearly does not determine, policy outcomes (see Marsh and Rhodes, 1992, p. 23). This is not to suggest that 'deals' are struck which *always* favour a particular group, rather access allows participation to a policy making environment which seeks to minimise political disturbance and establish (a favourable) continuity in policy. And yet, policy during this period was by no means 'out of tune with the times'.

Throughout western Europe in the 1970s the 'green issue' figured only at the margins of political debate and did not provoke the type of animated political exchanges synonymous with the 1990s. While publications such as Rachel Carson's *Silent Spring* (1965) and environmental disasters such as the Torrey Canyon oil spill (1967) had brought the 'green issue' to the attention of a wider audience, legislation did not materialise until the 1970s.[5] The USA led the way with the 1970 Clean Air Act, followed by Sweden, Germany and the UK. However, the period was marked by different national attempts to define an environmental policy. Across Europe, governments chose initially to follow the lead taken in the USA with the formation of scientific and technical advisory bodies (Weale, 1992). In Ireland a range of institutions already existed and simply subsumed this task. Institutional change, where it took place, varied. In Germany the preference was to adopt pollution control functions in an existing Ministry, whereas the UK (1970) and Ireland (1977) opted to establish a new Department of the Environment.

Differences also emerged on the style of policy. As Weale notes, where political systems favoured rule making in a formalised manner (USA and Germany) there was an explicit legal review of standards (Weale, 1992). In the UK and Ireland, administrative convention largely dictated that regulators be allowed discretion. Preference in Ireland and the UK, therefore, was generally accorded to establishing voluntary codes of practice and operating a pragmatic regulatory philosophy.

The Alkali Works Regulation Act (1906), which governed the control of air pollution in Ireland, typified many of these strictures. Apart from a few rudimentary emission standards prescribed in this Act and the Control of Atmospheric Pollution Regulations of 1970, there were no detailed emission standards operating in Ireland in the late 1970s (Scannell, 1982, p. 68). The legislation's limited scope was compounded further by the fact that it had not been updated to cover new production methods. Displaying a distinct penchant for the pragmatic, and perhaps a well-rehearsed capacity to improvise in difficult circumstances, local authorities opted to use the Local Government Planning Acts of 1963 and 1976 as an alternative mechanism for pollution control.

While the limitations of the Alkali Works Regulation Act revealed the political pragmatism which lay at the heart of the practice of local government, it also illustrated other weaknesses in the regulatory framework. Most notably, the lack of enforcement powers and an unwillingness to define precise environmental standards. The Alkali Works Regulation Act stipulated that works should be registered with the inspectorate and that the principle of 'best practicable means' would inform the standards for certification. The term best practicable means was never defined, but as Scannell points out, in practice it closely approximated to the expression 'all practicable measures' adopted in the Factories Act in the UK which entailed having regard to the state of knowledge at the time, particularly scientific knowledge (Scannell, 1982, p. 65). However, the inspectorates' powers of enforcement were severely curtailed under the legislation and, while changes to a particular operation could be recommended, certificates even refused, the option to close down a works was not available. And yet, as Scannell points out, a survey undertaken by the IDA showed that all three sulphuric manufacturing facilities in the country 'had given' or were 'still giving rise to problems at certain times' (cited in Scannell, 1982, p. 68).

The problem was not simply about weak or outdated legislation. Neither was it a case of negligent regulatory behaviour on the part of the inspectorate. Rather, the source of the difficulty lay in the inspectorate's need to have 'one eye' on the political repercussions of stringent regulation and the economic consequences of a works being closed. As Scannell succinctly observes, a move to close down an operation was an avenue only the 'politically insensitive would hazard' (Scannell, 1982, p. 68).

Weak legislation and a regulatory style intolerant to stringent enforcement was a recurring theme in environmental regulation in the late 1970s and early 1980s. Nowhere was this more evident than in the area of waste disposal, where local authorities were under no obligation to dispose of waste other than domestic refuse. And yet, in practice, both local authorities and private operators were allowed to dispose of waste in local authority dumps and, although local authorities were the responsible regulatory bodies for waste disposal, they did not employ personnel to enforce the legislation. Not surprisingly, prosecutions were few and far between (Scannell, 1990b).

The Fisheries Acts 1959–1962 and the Local Water Pollution Act of 1977 also typified many of these weaknesses. Organised largely on a voluntary basis, the Fisheries Acts 1959–1962 did not provide protection against ground water pollution from agricultural activity. Moreover, the 1977 Water Pollution Act offered a defence to the polluter on the grounds that prosecution would not take place provided that he or she had taken 'all reasonable care' to prevent entry. Although not confined to agricultural pollution the issue of the 'good defence' clause has always provoked considerable controversy in Irish political circles and the question of whether or not it should be removed from the legislation became closely connected with debates about how much influence the agricultural lobby should have on the Irish policy-making process. The Irish farming lobby and the Department of Agriculture, Fisheries and Forestry have always resisted attempts to remove this form of defence (Taylor, 1998). For legal purposes, the phrase 'all reasonable care' was usually equated with procedures which accorded with 'good agricultural practice'. However, what constituted 'good agricultural practice' was, of course, defined by the Department of Agriculture and not by those consigned to prevent water pollution. Altogether more disconcerting to environmentalists was the Inter-Departmental Environment Committee's comment that the Act did not provide the 'appropriate means' to deal with agricultural pollution resulting from 'fertiliser run off, chemical seed dressings, herbicides and insecticides, and silage making operations' (cited in Scannell, 1982, p. 105).

In many ways, given the well established links between the Department of Agriculture, Fisheries and Forestry and the farming lobby, agriculture became virtually self-regulatory. It was an issue upon which Scannell was moved to conclude that the most effective environmental controls for agriculture lay essentially outside of formal law and were generated in the grant-aided farm modernisation scheme organised by the Department of Agriculture (Scannell, 1990a).

The intention of such schemes was to provide an incentive for farmers to operate in a more environmentally friendly manner, thereby establishing greater uniformity in agri-environmental practices. Grant aid would be given only after environmental conditions attached to planning permission had been accepted. However, the guidelines were sufficiently flexible to allow significant deviation, and there was nothing to prevent a 'sufficiently motivated' farmer from avoiding constraints if he or she was willing to forego the grant aid (Scannell, 1982, p. 106).

By the late 1970s the practice of adopting environmental conditions in planning permission as a method for pollution control had become widespread. In the absence of coherent legislation, the appeal of this particular avenue clearly lay in the discretion available to planning authorities to attach conditions to planning applications. However, the financial constraints imposed upon local authorities meant that they did not possess the expertise, personnel or resources to carry out the tasks which were assigned to them. Advice, where necessary, was sought from either the Institute for Industrial Research and Standards (IIRS), the Department of the Environment or the private sector. It was, as the Inter-Departmental Environment Committee noted, a precarious practice and not without limitations. In its report of 1977 the committee concluded that the use of planning permission as a mechanism for environmental regulation was extremely limited and cumbersome since it could not easily be changed in the light of new circumstances (IDEC, 1977). However, while local authorities were keen to exploit the 'flexibility' planning permission offered as an instrument for pollution control, enforcement remained the exception rather than the rule. And yet, despite the difficulties encountered in the use of planning permission as a mechanism for pollution control, the practice continued into the 1980s. It is a fact all the more remarkable when we consider that section 26(2) of the 1963 Local Government (Planning and Development) Act does not contain a single reference to pollution control (Scannell, 1982).

The instances of lapses in the rigorous implementation of regulations highlight one of the most important features of policy during this period; the discretionary nature of much of the legislation. It was a problem not consigned exclusively to the regulatory role of local authorities. On the contrary, it appeared almost to be an endemic feature of governmental practice, where Ministers' positively balked at the prospect of advancing environmental regulation. It was not that there was an absence of sufficient legislative powers. The Local Government (Water Pollution) Act 1977, for example, empowered the Minister to set up a Water Pollution Advisory Council with a mandate to advise her or him on all aspects of water pollution and to submit an annual report in this area. The Minister also possessed the requisite powers to compel a local authority to create and implement a water quality management plan. It should surprise few of those conversant with environmental issues in Ireland during this period to find that no Minister ever undertook such a task. Furthermore, while the Act empowered the Minister to set standards for sewage discharge by local authorities, none were ever set (Taylor, 1998).

In public policy circles these are often referred to as 'implementation deficits' or, if you prefer, a discrepancy which occurs between legislative intent and regulatory practice. This contains at least four interconnected themes. It often involves a failure to consider the impact of pollution in one media on another (cross media transfer), the effect of pollution downstream, or the consequences of pollution which travels across nations and continents (transboundary pollution – acid rain). Consequently, policy tends to ignore the environmental dimension to a wide tranche of other public policies and thereby fail to reflect upon the problems which may lie ahead. Here, governments tend to concentrate upon highly visible forms of pollution where short-term 'quick fixes' gain important political kudos. In other words, these types of policy response are more suited to the exigencies of the electoral cycle. Short-term, ad hoc and reactive it lacks a more 'holistic' appreciation of the complexity of the environment. For example, dumping waste (as opposed to either recycling or reducing consumption) may offer a low cost option but can often ignore future problems in maintaining landfills and preventing the 'leaching' of hazardous substances. These implementation deficits were compounded by the fragmentation of responsibilities between different public bodies and a preference among inspectors to opt for negotiated compliance rather than enforce specific standards. It was a problem not unique to environmental policy but a feature common to other areas of public policy where amicable working relationships are viewed as beneficial to both sides of the regulatory framework. Here, it is important to recognise that operators of a particular production process, or a professional organisation, are often in possession of important knowledge from which to inform regulatory judgements (Weale, 1992, p. 16).

That discontent pervaded the environmental lobby over the state of Ireland's ecology during this period hardly requires comment. For many, however, potential salvation lay ahead in a tract of EU Directives emerging from Brussels. Looked upon idolatrously by those interested in the conservation of Ireland's green image, they were greeted as the harbinger of a new sense of environmental awareness in Ireland, stimulating change and providing a new impetus to reform. Such sentiments persisted until the late 1980s and were captured in F.J. McDonald's (1990) appropriately entitled essay, 'Can Europe Really Help?' Reflecting on the state of environmental regulation in Ireland, replete with its flaws in planning regulations, an absence of local democracy and 'rabid individualism', McDonald wistfully lamented that:

> sometimes I wonder, ... whether it is easier to change things in a totalitarian society than it is proving to be in this democracy of ours. But perhaps, with the help of friends in Europe, we will see the light some day (McDonald, 1990, p. 111).

And yet, it is often forgotten that during this period environmental policy at EU level had achieved nowhere near the prominence it now enjoys, evidenced in the fact that its origins lay not in the protection of the environment or any widespread recognition for more stringent environmental standards, but in the imperative to reduce trade distortions and avoid the possibility that different (and possibly conflicting) regulatory standards would become an obstacle to business competition (Lowe and Ward, 1998). It became a distinct policy sector only in the 1980s.

A succession of Directives during this period strengthened the Community's influence on environmental policy and by the mid-1980s it had become the fastest growing area of EU policy. While Chernobyl may have been the single most important catalyst to this shift in priorities, there were other notable environmental problems which spurred change; acid rain, global warming and the shipment of toxic waste. In addition, the election of Green MEPs in the mid-1980s ensured both the presence of a green environmental lobby in the corridors of power in Brussels and a focal point for green protest. In Ireland, the response to such developments was mixed. While legislation was forthcoming, it was, as Scannell observes, often flawed. Government tended to incorporate EC Directives into Irish law, a move which may have eased the passage of legislation but often left it ineffectual (Scannell, 1990a). The more common practise was to use regulations under the 1972 EEC Act, which meant that only summary offences for pollution carried a maximum penalty of £1000.

The underlying logic of a Directive is to allow a Member State the flexibility to incorporate European Community law into their own legal regime in a way that best fits the national situation in terms of ease of application and enforcement (Schaefer, 1991, p. 110). In the Irish case this is not generally achieved. The Directives are usually not modified to fit the particulars of the Irish situation at all; since (it) is more usual for them to be translated verbatim into Irish law (Coyle, 1994).

On a more positive note, EC Directives shifted attention away from pollution control to establishing precise environmental standards, a feature of regulation in Ireland which had been conspicuous only by its absence. This was most evident in the water pollution Directives which established a scientific basis for policy. The success of this initiative can be gauged from the fact that while few of us are aware of the technical requirements that need to be met, we are all familiar with the importance of the blue flag.

While Directives are binding – that is governments must achieve designated targets – the choice of methods by which these results are achieved varies considerably across Member States. This has had an important bearing upon the impact of the Directives on regulation in Ireland and, in many ways, has militated against sustained improvement. It was common practice, for example, for Departmental circulars to be issued on foot of new Directives, a move which attracted criticism from within the EC because the details of these circulars were never made public. As such, the potential for individuals or concerned groups to challenge regulators or polluters was significantly curtailed. Moreover, in a number of important instances, most notably groundwater pollution, the implementation of these Directives was defective.

The decision by the Department of the Environment not to enact new legislation in this field, for example, meant that while private sector operators could be regulated under the 1977 Local Government and Water Pollution Act, no such provision existed to prevent public authorities from directly discharging List 1 dangerous substances into groundwater (Scannell, 1990b). These are not minor discrepancies, or the idiosyncratic observations of someone fascinated by the complex minutiae of environmental law, but raise issues which can have important environmental consequences. This is demonstrated in the problems which arose because the DoE felt that Directive 76/160/EEC was adequately catered for in the Local Government (Water Pollution) regulations, 1978. However, as Scannell points out, local authorities are exempt from the requirement to obtain a licence to discharge trade effluents, a situation which contradicts Article 3 of the Directive. It was an issue which emerged in a controversial application from Sandoz in Cork which sought authorisation to discharge effluents to a local authority sewer. While they required a licence for this, there was no provision for public participation. Neither did the public have recourse to challenge the quantity or quality of the discharge. However, had Sandoz wished to discharge the effluent into the sea, both appeals and challenges to the licence could have been made (Scannell, 1990a, pp 88-9).

The problems which become manifest over the translation of EC Directives were complicated further by the fact that much of EC legislation often allows scope for significant variation to take place in Member States. Thus, for example, in the case of the 1977 Directive on the sulphur content of certain types of oil, Ireland was able to extract a five year exemption from having to comply with the second stage of the programme. In addition, it also secured a five year 'breathing space' on implementing the EC Directive to reduce the lead content in fuels. On both counts, the Department of the Environment defended its opposition to the Directive on the grounds that the investment required to convert Whitegate oil refinery would have placed the operation in 'jeopardy' (Scannell, 1982, p. 74).

More disconcerting to the environmental lobby was the fact that difficulties were not consigned simply to the translation of European Directives. Problems existed elsewhere, largely as a result of the fragmented and discretionary character of legislative responsibilities and a failure to define precise environmental standards. As a consequence what finally transpired in the late 1980s was an emerging crisis in environmental regulation as the Irish polity experienced an upsurge in environmental protest. There were at least three principal actors involved in this unfolding ecological drama which deserve attention; local authorities, the business and agricultural lobbies.

With regard to the role of local authorities it is difficult to ignore the fact that a significant number of problems which lay at their doorstep could be traced back to the failure on the part of central government to sustain the necessary resources required to implement the explosion in Environmental Directives emanating from the EC. This was undoubtedly compounded by the fact that from the mid-

1970s onwards environmental policy debate has been conducted in an increasingly 'scientific and technical' manner. It is not simply that division exists over scientific explanation of the source(s) of pollution and their subsequent effect, a situation often downplayed in the scientific community, but that it often makes the choice of the instruments of environmental policy all the more contentious. Defining a problem, and then framing a response, becomes an increasingly complicated (and costly) exercise.

During the 1980s, a period in which there was intense circumspection over the precarious state of public finances, it was hardly surprising that many local authorities found themselves having to rely on outside expertise in complex areas; waste management to the private sector and air quality monitoring to the health boards. More importantly perhaps, given the regulatory function of local authorities, they did not possess the scientific expertise to comprehend fully the technical implications of the control measures for which they were responsible. As Leech has observed:

> When the range of expertise required is considered, covering the broad area of the aquatic environment, toxicity, eco-toxicology, assimilative capacity, standards, control technology, air pollutants, acid rain, toxic wastes and chemicals, air and water quality modelling etc., it is evident that few local authorities employ or *could afford to employ* the broad range of expertise required (Leech, 1989, p. 412, emphasis added).

In a similar vein, Keohane cites the case of one local authority environmental officer who drew up planning permission for Penn Chemicals in 1975. The company, which was expanding the range of products to be produced at the plant, got an extension to the original planning permission. However, no additional air pollution control conditions were added, which meant that either the environmental officer involved was not aware that methyl mercaptan was going to be a by-product of the new process or he was not aware of the environmentally deleterious properties of methyl mercaptan. Either way, a serious lack of expertise was demonstrated (Keohane, 1987; Taylor, 1998).

A further problem for local authorities was the contradictory nature of their regulatory function. On the one hand, they were charged with monitoring and regulating the environment, and yet on the other hand, they were in competition with one another to court the investment of multi-national companies. Almost inevitably, the determination to enforce stringent environmental standards was compromised by their role as development corporations, a scenario complicated further by the fact that they were exempt from many of the pollution controls they were supposed to be enforcing on others (Scannell, 1990b). For example, Scannell notes that the 1979 waste regulations implementing EC directive 75/442 do not refer to the obligations on the state authorities to promote re-cycling, carry out periodic checks on waste disposal facilities, to respect the polluter pays principle and to forward regular reports to the Commission. Equally, the toxic waste regulations of 1982 implementing EC Directive 78/319 do not refer to the requirement for all bodies (public or private) storing, treating or depositing toxic and dangerous waste to obtain a permit from the appropriate body (Scannell, 1990a).

The contradictory nature of this regulatory function intensified during the 1980s amid concern that further enforcement of pollution regulation in the multi-national sector could de-stabilise Ireland's 'climate for investment' (Allen and Jones, 1990; Leonard 1988). Leonard, for example, highlights the importance of the threat of plant closure in the Irish case. The Chief Planning Officer in County Waterford, in an interview with Leonard, stated that there must:

> be a manual that US industrialists use to deal with local authorities when they want to get the plant to clean up the muck. First make some veiled threats to close. Then get the workers and the unions involved.

> Then just sit back and let concern about jobs and the economy divide public opinion enough to immobilise local officials (Leonard, 1988, p. 187).

The incongruous role of public bodies acting as both gamekeeper and poacher was replicated at a national level where the Department of the Environment's portfolio extended beyond environmental regulation to include responsibilities for housing, roads, sanitary services and local authority functions. Such an extended set of responsibilities necessarily brought it into a situation which conflicted with the overall direction of environmental regulation (Taylor, 1998).

In the wider economic sphere the Industrial Development Authority (IDA), which played a critical role in Ireland's quest for economic modernisation, also attracted the attention of environmentalists. Its function was to encourage and co-ordinate investment projects and, during the 1960s, 1970s and 1980s, virtually had a free hand with industrial development, providing that a steady flow of job announcements materialised (Lee, 1990, p. 473). Its original concern lay with attracting labour intensive industries, thereby offsetting high levels of unemployment. However, by the late 1970s the IDA changed tack and sought to encourage industry which would have a high value added content, providing employment for skilled labour and trained technicians. It had recognised both the threat posed by cheap labour markets in the third world and the importance of shifting the focus of exports to European markets. Consequently, its gaze fell upon the multi-national chemical industry, and in particular the pharmaceutical sector (Taylor, 1998; Leonard, 1988).

The problem, as far as the environmental lobby was concerned, was that many of the IDA's clients were companies whose environmental record was less than favourable. To such critics, it seemed that the history and concentration of chemical companies in Ireland, particularly in the area of Cork Harbour, was simply not consistent with the charge that environmental considerations are negligible *vis-a-vis* choosing an investment location (Mullally, 1993). Put simply, as far as some elements of the environmental lobby were concerned, the IDA operated a 'cosy cartel', in which its principal function appeared to be little more than applying strong political leverage on local authorities.

In part, this can be explained by the fact that a significant factor in the IDA's success during the 1980s was the one stop service it provided, which meant that multi-nationals entered negotiations with a single organisation, avoiding the complications of overlapping agendas, parallel power structures and bureaucratic inertia (Barry and Jackson, 1989). The temptation to conclude that close ties – almost an endemic feature to the successful completion of its task – conceal more sinister motivations is one which should be strenuously avoided. There is a need to caution against promulgating the simplistic view that the IDA was involved in some form of duplicitous act designed to conceal the environmental impact of new investment. All too often such arguments conveniently ignore either the changing political terrain upon which the IDA operated or believe that it acted in some political vacuum, unaware of the political pressures arising from increased unemployment and emigration.

This is not to suggest that the IDA was not averse to 'impressing' upon local authorities the need to relax environmental controls in order to secure 'a competitive advantage over other "potential" locations' (Leonard, 1988, p. 127). Neither was it the case that the agency was unaware of the environmental impact of some of its companies since, as one official remarked, there wasn't much planning in the 1970s; 'we just assumed that we would have to take whatever industry we could get and that dear old Ireland would have to make some trade-offs' (Leonard, 1988, p. 131). For some, this is a phrase which confirms the 'fact' that the agency was fully aware of its approach and was at pains to keep it off the record. However, attention should surely focus on the absence of 'planning' and the recognition of the need to make 'trade offs'. As in so many other areas of Irish politics, policy was informed by a style which was essentially conservative, ad hoc, reactive and incremental. It seems plausible to suggest that on this matter at least, that it was also 'off the cuff', since there was little in the way of previous policy to adapt, a matter which clearly made 'life' more difficult for officials working in this area.

If we are to understand the 'dynamic nature' of environmental policy during this formative period then it remains crucial we recognise that environmental considerations were just one of a number of competing concerns which shaped the terrain upon which the IDA acted. Of course, the agency had close ties with multi-national companies – its function after all was to assist in the creation of opportunities, 'smooth over difficulties' and, if necessary, construct a defence against environmental protest. But we need to consider that political debate (and policy) were shaped by a number of different (and conflicting) actors on issues as disparate as tax, welfare, unemployment and emigration. As such, environmental policy and its deficiencies during this period can be understood only in the dynamic interplay between the IDA, government, local authorities and the business and agricultural lobbies.

Here, this chapter suggests that the IDA was an important participant in a 'closed' industrial policy community populated by other prominent political actors such as the Federation of Irish Chemical Industries (FICI), the Irish Business and Employers Confederation (IBEC) and the Institute for Industrial Research and Standards (IIRS).[6] Amid increasing public disquiet about the environmental record of some of its multi-national clients during the late 1980s (most controversially Merrell Dow) the challenge for this policy community was to present an 'alternative vision' of regulation to that offered by the increasingly politicised environmental lobby. In contrast to the conspiratorial tone of its critics, therefore, this chapter suggests that the IDA was engaged in a far more subtle and complicated process of engineering consent around new levels of 'acceptable pollution'. It was a task in which the Institute for Industrial Research and Standards (IIRS) performed a key role in presenting a scientific and technical defence to the policy community's interests against opposition from environmental pressure groups (Taylor, 1998).

The importance of bodies such as the IIRS should not be underestimated. In a period of political flux in which contentious projects became more commonplace, the role of the IIRS became crucial to the policy community's attempt ward off the challenge of the environmental lobby. The role of the institute was invaluable if an aura of scientific credibility was to surround determinations on environmental issues. Remember, these were not absolute standards cast in stone. In practice, environmental permission values varied according to whether it was: a new or existing facility (because more stringent regulation may have threatened the economic viability of an older plant); the capacity of the environment to absorb pollution or, where a sector's lobbying capacity was able to extract concessions or 'resist' more stringent enforcement. It was hardly surprising that as far as the environmental lobby was concerned, the primary function of the IIRS was to provide 'a scientific and technical veneer to state projects, designed to diffuse anger and fear' (Allen and Jones, 1990, p. 259). The important feature of this policy community was not so much who was in it as who was not. Despite assurances throughout the 1980s that the Irish Government recognised the need to embrace the 'green issue' there were no environmental lobbying organisations or representatives in the industrial policy community.

This was a situation replicated in agriculture where, despite the overwhelming evidence of the detrimental impact of agriculture on the environment, there were no 'green representatives' in agricultural policy formation. It was a situation which served to reinforce a remarkable level of continuity in policy, favouring initiatives which were discretionary and predominantly orchestrated to subsidise production rather than reconstruct the relationship between agriculture and the environment.

It was not that the Department of Agriculture was reluctant to avoid a powerful lobbying group such as the IFA, although whether the political will existed is clearly debatable, rather that it was predisposed to negotiate change, an approach reinforced by a belief that little, if anything, could be achieved through a protracted struggle over principles. Better to seek grounds for compromise, utilise the knowledge of the IFA and secure compliance on incremental change.

As such, the IFA's access to the policy making environment, at both a national and EC level, ensured that policy was formulated around the use of discretionary incentives to increase production, a move which had a largely unanticipated and significantly deleterious effect upon the environment. From the farm modernisation scheme onwards, the environmental component to agri-environmental policies was subsumed in negotiations designed primarily to maintain income transfers to farmers. Political access was therefore vital to ensure that any new schemes were replete with provisos; from (voluntary) participation to inspection levels. It was, above all else, imperative that new schemes consider fully the economic impact of any environmental dimension to future agricultural practice.

By the mid to late 1980s the level of opposition in both Ireland and Europe to this system had become formidable as a range of political groups argued that CAP was in need of serious reform (Regan, 1994). Both wildlife habitats and water quality were impaired as intensive agriculture increased its use of fertilisers and pesticides. The response from within the agricultural lobby to calls for reform was to reconstruct the farmer as both producer and custodian of the countryside (Regan, 1994, p. 75).

In seeking to make 'accommodations to reality', then, Ireland's environmental policy style was predisposed to stress the importance of making concessions. Incremental, ad hoc and largely reactive it was an administrative style which articulated the need to avoid confrontation with organised interests and eschew radical reform. By the late 1980s, however, it had become clear that environmental regulation was effectively hamstrung by inadequate enforcement and a lack of coherence, flaws which ultimately precipitated a 'crisis of confidence' in the state's determination to protect the environment. In 1987, for example, Merrell Dow announced plans to set up a plant in Killeagh in the Womanagh Valley in County Cork. The residents in the area objected to the location, primarily on the basis that a pharmaceutical plant should be located in an industrial estate as opposed to the rural setting of Killeagh. However, the company got planning permission despite all the local opposition. The local authority decision was appealed and following an oral hearing, An Bord Pleanala upheld the decision, albeit with a number of conditions. A case was taken to the High Court on the jurisdiction of An Bord Pleanala to make a decision on the matter in the light of alleged deficiencies in the information submitted in the planning application. The High Court ruled that An Bord Pleanala did indeed have the power to make the decision in this case and the court refused to overturn the decision of the planning authorities. In the end, after nearly three years of controversy and amid calls on the Minister for the Environment to intervene, the company announced their decision to pull out of Ireland altogether. They cited a merger with another company and a consequent need to re-evaluate their production needs as the reason. Many feel that local opposition to the project was a more credible reason.

Suffice it to say, the case was highly publicised and served to focus the attention of the public on the arguments of those who saw Ireland's environment threatened by the forces of development. The prevailing mood of this period was succinctly captured in an editorial of the *Irish Times* which declared that:

> there is a growing conviction ... that the provision for policing environmental controls are less than adequate, and that the industries are, in effect, left to police themselves (*Irish Times*, 5 September 1989).

The challenge presented by community and environmental groups had served to question the influence of the industrial and agricultural business lobbies. Moreover, the extent of cross party support for change indicated the pressing political need to replace a myriad of confusing legislative arrangements with a single agency to provide a co-ordinated, integrated and fully transparent approach to environmental regulation. Indeed, the very structure of that agency, the EPA, its obligations, operating rationale and regulatory philosophy bear full witness to the scars of a struggle over the increasingly discredited environmental policy regime of the 1980s.

'A Watchdog, not a Bloodhound':
The Formation, Structure and Regulatory Ethos of the Environmental Protection Agency[7]

In the summer of 1989 the Fianna Fáil and Progressive Democrat coalition announced that it was to replace a confusing medley of environmental regulations with a single agency responsible for protecting Ireland's environment (*Irish Times*, 6 September 1989). The announcement was welcomed on the opposition benches and, perhaps more importantly, endorsed fully by an array of environmental pressure groups. Indeed, Emer Colleran, Chairperson of An Taisce and an activist later involved in the environmental dispute over the proposed location of an interpretative centre at Mullaghmore, was moved to declare that the government's programme was 'light years away from what we have had so far' (E Colleran, *Irish Times*, 13 July 1989).[8] It also struck a chord in the Workers Party, which had called for change along these lines in its policy document *The Environment is Where We Live* (*Irish Times*, 9 May 1989). In a similar fashion, both the Labour Party, *Proposals for a Socialist Environment*, and Fine Gael, *A Clean Environment*, had joined the chorus for a new environmental agency (*Irish Times*, 6 June 1989).

A theme which pervaded much of the debate during the passage of the legislation was that the EPA would finally lay to rest the enduring conundrum of reconciling economic growth with environmental conservation. It should come as little surprise then that amid the clamour to celebrate its arrival the EPA should be heralded as a new beginning in Irish politics, that finally the 'green issue' had been accorded its rightful place on the political agenda. No longer located at the periphery of political dialogue, arguments about the ascendancy of the green issue in Irish politics also found resonance in moves to identify elements of an new, embryonic modernity in Ireland. As an ecological cypher of these 'new times', it also lent itself easily to those who proffer that it represented the institutional expression of a new, cosmopolitan European State, confident in its capacity to retain a green image while competing in the new global economy (Culliton, 1992).

While precise definitions of this ecological modernity have proven elusive it is perhaps plausible to suggest that they find common ground in the belief 'that countries may undertake the great leap forward over the phase of "dirty" industrialisation into the fully ecologically modern condition' (Christoff, 1996, pp 487–488). As with most umbrella terms there are a number of competing views about what constitutes ecological modernisation (EM). For Jänicke, it is essentially a response by business to the emerging pressures of the global economy which accords a priority to technical cost-minimisation. Improvements in environmental standards may emerge, but its principal aim is to sustain competitiveness. It does not entail changes in cultural or political values but is almost an 'afterthought', as business responds to changing tax regimes, consumer taste and production methods (Christoff, 1996, p. 480). Hajer (1995) on the other hand, uses the term in a broader sense, seeking to explain changes which have taken place in environmental policy making and believes that change has been underway in several areas: a policy preoccupied with 'react and cure' is more likely to be concerned with anticipation and prevention. In addition, at the macro-level there is no longer the feeling that the environment is somehow a 'free good' but an invaluable public resource. At the micro-level there is greater credence given to the argument that pollution prevention pays and a shift toward more participatory regulatory regimes which 'seek to bring to an end the sharp antagonistic debates between the state and the environmental movement' has taken place (Hajer,

1995, pp 24–30). As with Jänicke, there is an important economic dimension here, pollution prevention *must* pay.

For Weale (1992) EM is seen as a policy discourse used to (re)organise the state's response to ecological dissent and forms the basis around which a new understanding of environmental conservation is constructed in order to sustain economic development. Once the old paradigm of a conflict between the environment and the economy is challenged, then a transformation can take place in which there is an ecological dimension to all relations. The dichotomy between business and ecology is replaced by the long-term (ecologically aware, progressive business) and those engaged in short-term profit taking. The state's role is to provide support for ecologically aware producers/consumers to influence cultural and political behaviour (Weale, 1992).[9] It is a set of arguments which contain at least three themes of relevance to our discussion. First, the intimation that a new complementarity between ecology and the environment is constructed. Second, that it involves a transformation in pollution-control technology and, in particular, a shift from a preoccupation with abatement or 'end-of-pipe' solutions to prevention. Finally, it endorses a reorganisation of administrative conventions and regulatory practices with integration to the fore. In general terms, debates on this issue coalesce around the nature of state or market failure in previous environmental policy paradigms.

Within the tradition of environmental welfare economics it is argued that markets 'fail' because environmental resources have not been properly valued (in other words they are seen as free and therefore inappropriately used). As such, they recommend the adoption of pollution taxes so that polluting industry will 'internalise' or consider the cost/benefits to be gained from further pollution. This should encourage industry to adopt more eco-friendly production practices in order to achieve an optimal balance between productions costs and pollution (see Helm and Pearce, 1991 or Buchanan, J and G. Tullock, 1975). It is an approach with problems. Most notably the reluctance in many Finance Departments (including Ireland) to entertain the idea of 'earmarked' taxes. Notwithstanding this dilemma, there is also the 'delicate task' of determining the level at which the tax should be set, a precarious enough exercise without the lobbying pressure likely to emerge from powerful, polluting multi-nationals (Eckersly, 1995).

A further, more radical line is taken by free market environmentalists who suggest that the source of the problem lies in state-led regulation (it's bureaucratic, cumbersome and open to 'persuasion' from powerful lobby groups). The solution, in crude terms, lies in privatising the environment. The argument runs that once property rights are established, people will have a vested interest in protecting their assets, possess greater knowledge of environmental requirements and would not be liable to regulatory capture (this occurs when monitoring and enforcement are compromised by cosy relations between inspectors, regulators and those being regulated). The strength of this approach undoubtedly lies in its critique of bureaucratic failure rather than its idiosyncratic recommendation to 'privatise the oceans'.

The theme of bureaucratic failure is one which has attracted the attention of those such as Jänicke (1990) and Dryzek (1987). According to Jänicke, the problems in environmental policy lie not simply in regulatory capture, or the failure of pressure groups to exert their influence in decision making, but in a breakdown in communication between those who negotiate and decide on policy (bureaucrats) and politicians. The result is that policy is influenced by the bureaucratic need to look for solutions that favour political (electoral) success which demands that the interests of business are taken into consideration (jobs = votes). Policies therefore tend to be predictable, routine and standardised band aid solutions (Eckersly, 1995). They are not flexible, integrated, proactive or long-term (see Jänicke, 1990).

At a bare minimum, this new ecological condition demands the rejection of an approach which simply transferred the problem of pollution to another place, media, or time. This was perhaps most visible in the billowing smokestacks dotted around the industrial landscape of Europe, designed to allow the dispersal of pollution to higher levels of the atmosphere so that 'toxic' substances could be diluted to an 'acceptable' level. It was a limited, short-sighted and largely *reactive* approach, which sought to alleviate rather than prevent pollution. It was also a policy paradigm in which technical solutions tended to concentrate upon removing, transferring or abating pollution excess. In contrast to such 'end of pipe' solutions, the new ecological condition perceives pollution (and its response) to be part of the production cycle as a whole, not an isolated feature or by product. In other words, it is a more *proactive* approach which examines the 'life cycle' of the product and how new production practices, waste reduction techniques and technology can prevent pollution.

That this new ecological condition envisages a new complementarity between the environment and the economy is confirmed in the rejection of the idea that stringent environmental regulation necessarily retards economic growth. Grabodsky, for example, notes that it is no accident that the world leaders in the production of pollution control technologies are OECD countries with the most stringent environmental regulations in the world. Germany leads in water pollution abatement, Japan in air pollution control, the Netherlands in soil remediation and the USA in toxic waste disposal (Grabodsky, 1995, p. 204). On the contrary, for those such as Wiche (1984), the new ecological condition recognises that developments in environmental protection (either through new technology, production practices or products) become a commodity with a potential market.[10]

It may be difficult for us to assimilate the progress which has occurred in pollution control technology, but we are all familiar with the changes which have taken place in supermarkets, where the introduction of environmentally friendly products, from toiletries and detergents to organic foods, has substantially increased. Grabodsky remains optimistic about the potential for this demand-led regulation to initiate change. In other words, changes in our consumption patterns (our choice of eco-friendly detergents) offer a solution to the problems of state-led regulation identified by those such as Jänicke. In his opinion, many industrialists feel that consumer demand is more stringent than that of regulators, citing the views of one Swedish pulp and paper manufacturer who insisted that 'it would be easy if we only had to cope with the regulators, it is the consumers pressure that changes us most' (Grabodsky, 1996, p. 205). I am far more circumspect on this issue. For consumers to make 'rational choices', choices upon which they can make a 'green' judgement, it would require (at the very least) far more information to be contained in product labelling. It seems this issue has been around an awful long time and little so far has been done. Indeed, eco-labelling was part of the EPA's remit and yet, almost nine years down the road, we seemed to have progressed very little. Ask yourself, which washing powder do you buy and are you aware of the substance that is the principal source of damage? How do I look for the alternative? And that is only one product.

In the field of regulatory regimes, EM anticipates a departure from an administrative system in which multiple permitting was required and where enforcement was the responsibility of diverse (and often conflicting) bodies. Under such circumstances, it was commonplace for companies to undergo several inspections for a variety of emissions with little or no co-ordination. Time consuming, cumbersome and overly bureaucratic, it is clearly inappropriate to the prevailing conditions of the global economy where the new mantra of business is flexibility. In contrast, the new ecological condition anticipates pollution problems, elevates the importance of prevention and favours the adoption of a regime based on integrated pollution control (IPC).[11] Where hitherto fragmentation and division were the order of the day, co-ordination, integration and transparency should now reign.

It is an alluring cocktail of science, economics and administrative change. Certainly, given the change which has taken place in Ireland over the last decade (not least in environmental regulation) it should surprise few that the 1992 EPA Act should be greeted as evidence of an emerging ecological modernisation, containing as it does many of the central principles of the doctrine; transparency, IPC licensing, a shift from abatement to prevention and the adoption of on-going improvements to environmental standards. Indeed, the agency envisages that IPC:

> would eliminate or minimise the risk of harm to the environment by preventing the emission of potentially polluting substances wherever it is practicable or to minimise such emissions where it is not practicable (EPA, 1996, p. 2).

The EPA Act itself is more circumspect and avoids using the term 'harm' largely because senior civil servants were troubled by the possibility that it may lead to confusing interpretations in the courts. This contrasts with legislation in the UK, where the Environmental Protection Act, 1990 stated that:

> pollution of the environment means pollution of the environment due to the release (into any environmental medium) from any process of substances which are capable of causing harm to man or any other living organism supported by the environment (Environmental Protection Act, 1990, ch 43, p. 2152; see also section I of the Irish EPA Act, p. 10).

The use of the phrase 'living organism' is an interesting one, particularly in the light of the recent furore over genetically modified crop trials involving Monsanto. Much of the objections to such genetic trials hinge upon the potential for harm as a result of genetic crossover or the debilitating consequences of large tracts of land with no weeds to support insects and birds.

It would be churlish to suggest that positive developments have not been achieved or improvements sustained. The new regulatory powers available to the EPA, the development of new monitoring techniques and a welcome zeal to allow access to environmental information surely represent more than simply a 'political gesture' to an increasingly politicised environmental lobby. And yet, such progress should not be allowed to conceal flaws which remain. In particular, the legislative framework is replete with ministerial discretions and opt out clauses, features which sit comfortably within the Irish style of policy making. Indeed, it is the persistence of this soft regulatory ethos which remains the hallmark of this new framework, revealing rather more in the way of continuity than change with its discredited predecessor.

Above all else, the restoration of order was essential. The accent was firmly upon imposing uniformity in licence conditions, compressing the timescale for authorisations to be granted and 'calling time' on vexatious appeals. To a large extent this can be attributed to the fact that the origins of the EPA lie not in any serious attempt to overhaul administrative conventions or policy practices, or to an enhanced level of green awareness among politicians, rather it was constructed in order to reconcile the criticisms of the environmental lobby without threatening the free market ethos which has underpinned the Emerald Tiger.

Integrated Pollution Control Licensing

The EPA's remit was to be ambitious, befitting an agency fashioned to embrace more positively the green issue. Under the 1992 Act, the EPA was given a number of statutory functions, the most important of which was the introduction of integrated pollution control licensing (IPC). The intention was that IPC would supplant the practice of multiple permitting, serving to compress the application procedure and condense a range of licences (and responsible bodies) into a single pollution control licence. In addition, the EPA would also play an active role in the promotion of environmentally sound practices generally through a wide range of advisory, support and supervisory functions, provide guidelines and backup to local authorities, co-ordinate environmental monitoring, publish reports on the state of the environment and undertake environmental research. Among a wider audience, it was the agency's punitive powers which attracted the headlines, *apparently* confirming in legislation the government's pledge that it would no longer tolerate damage to Ireland's 'green image'. Offences under the 1992 Act, for example, carry penalties which are markedly more severe than previous environmental legislation. A maximum of £1000 and/or six months imprisonment accompanies a summary conviction while a conviction on indictment *could* lead to a maximum of £10 million and/or imprisonment for 10 years (EPA Act, 1992, p. 14).

The concept of integrated pollution has been growing in importance within the European community in recent years. The Fifth Environment Action programme of the Community, for example, designated it as a 'priority field of action'. In particular, the programme suggested that IPC licensing, when linked to improvements in the management and control of production processes, would give a 'new sense of direction and thrust to the environment/industrial policy interface' (EPA, 1996, p. 2).

For many observers, it was a 'rational' policy response which conceded the failure of previous administrations to fully comprehend the complexity of the ecosphere. Traditional policy initiatives, circumscribed by the archaic and fragmented character of Departments, considered emissions to air, water and land separately. It ignored the integrated nature of pollution which demanded a co-ordinated institutional response. Within such narratives, publications such as the *World Commission on Environment and Development* (1987) are often cited as examples of how an enlightened awareness in policy circles emerged, avoiding the lapses associated with previous forms of pollution control. It is a persuasive line of thought, the 'pieces to the jigsaw' all seem to be there and, perhaps more importantly, it fits with our intuitive feel of how policy 'should be made'. The narrow mandates of individual Government Departments and the closed nature of decision making would be supplanted by an integrated and transparent system of pollution control. A scenario in which a better scientific understanding feeds through to administrative and regulatory change. Appealing though this seems, it ultimately fails to consider the political dynamic which exists at the EU level, that despite certain tendencies toward supranationalism, 'political wrangles' between individual nation states still have an important bearing upon policy (Lowe and Ward, 1998, p. 3).

From the early 1990s environmental policy at the EU level has been moving progressively away from the style of regulation where legislation is binding. The British administrative style, with its preoccupation on costs, practicalities and the standardisation of monitoring and compliance procedures, has enjoyed greater prominence in the 1990s as the Commission became more conscious of the need to elevate 'British pragmatism' at the expense of 'continental abstraction' (Lowe and Ward, 1998, p. 18). This contrasted sharply with events in the 1980s where Britain contributed little to EC environmental policy. As Haigh remarked:

the occasions when British legislation ... has shaped Community legislation are far fewer than might have been expected of a country with a well established environmental policy (Haigh, 1984, cited in Lowe and Ward, 1998, p. 19).

Mounting concern over the poor record of implementing existing legislation, and a lack of administrative capacity in some Member States, meant that the EC was reticent about introducing a further tranche of environmental legislation. Consequently, its attention was redirected to the problem of implementation (Collins and Earnshaw, 1992, p. 214). Indeed, the late 1980s witnessed an escalation in infringement proceedings against Member States for partial compliance, non notification and the poor application of EC legislation. Ireland's position at the this point was considered to be 'middle of the road', better than the worst offenders such as Spain, France, Italy and the UK, but worse than Denmark, Portugal and Luxembourg. To a large extent this had been caused by the range and complexity of existing law in Member States and the differences in the power structures operating at national and sub-national level - whether local governments, or Lander in the German case, have substantial autonomy (Collins and Earnshaw, 1992, p. 225).

Such difficulties have not been assisted by deficiencies which exist within the institutional structures of the EC, where the failure to provide a sound scientific and legal basis to policy and ensure adequate consultation during the drafting of the legislation has compromised many policy initiatives (Collins and Earnshaw, 1992, p. 225). On this latter issue the House of Lords Select Committee of the European Communities observed that:

> just ten officials provide general advice on legislative drafting as well as being responsible for monitoring implementation in 12 Member States, comprising 15 or 16 different legal systems, in nine languages (Collins and Earnshaw, 1992, p. 225).

This has been compounded by the EC's decision making methods where, in theory, decisions in areas of the EC's exclusive authority were meant to be by majority voting. However, in practice, it has often required a unanimous decision. As Van Der Straaten illustrates, all that was required to unravel the momentum of change was the objection of a single Member State and, consequently, public decision making became a 'paradise' for private lobbyists seeking to persuade 'weak' governments to oppose legislation (Van Der Straaten, 1992, p. 65). It was hardly surprising then that consolidation became the maxim of the 1990s.

A further factor which contributed to visible change in style at the EU level was the persistence of the British government's position *vis a vis* subsidiarity, effectively eroding moves toward supranationalism or legislation which would be binding throughout the EC. Both the Single European Act and the Maastricht Treaty accentuated this penchant for greater pragmatism, reflected in a preference for framework Directives, soft law and voluntary codes of practice (Lowe and Ward, 1998, p. 24). In many ways this culminated in the Integrated Pollution Prevention and Control Directive (IPPC) (96/61/EC) which confirmed both Britain's increasing influence on environmental legislation and the importance of political bargaining in shaping final policy outcomes.

Implemented in 1990, the IPPC system reflects the British approach to tailoring controls to specific sites, with consideration given to the ability of the environment to absorb pollution (Sharp, 1998, p. 47). In contrast to events in the 1980s, the British government was determined to be involved in the minutiae of EC policy construction so that when the Commission indicated that it was to pursue new legislation in the early 1990s, the British succeeded in seconding a senior civil servant to work on the drafting of the legislation. The Directive which finally emerged was well received in the UK, resembling very closely its existing legislative framework. As Sharp (a retired British civil servant) noted, a number of European States (Germany and the Netherlands) tended to regulate emissions to each environmental medium separately. While the British proposal sought integration, it also asserted greater emphasis to a site-specific approach which was expressly designed to avoid different national standards emerging, and thereby restricting free competition between Member States. After a series of 'unstable negotiations', a delicately balanced compromise was reached which largely adhered to the British option. It was, as Clappison, a Junior Environment Minister pointed out, a decision which gave:

> the United Kingdom a complete victory on those major points of principle. The draft directive sets out a framework ... under which a site specific approach will be used, which takes account of the environment as a whole, and *adequate prominence will be given to consideration of the cost effectiveness and economic feasibility*. That framework is fully consonant with UK environmental policy ...
> (Clappison, cited in Sharp, 1998, p. 48, emphasis added).

Although the final Directive was largely modelled on the UK legislation it did contain some broader requirements, such as the inclusion of intensive farming. It did not, however, unduly concern British civil servants, who were acutely aware that there is 'inevitably some trading on detail necessary if a major policy gain is to be achieved' (Sharp, 1998, p. 48). A further reason for the delay in agreement over IPPC was the struggle to find a balance between BAT (reducing pollution at source) and environmental quality standards based on the principle that the cleaner the environment the less stringent the standard that will be required. The proposals allowed for the introduction of BAT, granting discretions to water down the proposals for some Member States. Germany was opposed and Finland demanded the right to implement more stringent legislation (see Maguire, 1995).

For the purposes of our discussion here the legislative passage of this directive is revealing, and shows that integrated pollution control had rather less to do with the 'rational' development of policy and rather more to do with the negotiations and political intrigue which form the backdrop to a struggle between competing policy styles at the European level. Moreover, it reinforces the view that IPC licensing was informed rather more by a pragmatic, co-operative and soft regulatory ethos, than some conversion to the principles of a new ecological condition. A desire to reinforce the discretion of Member States on whether to pursue further environmental regulation and a commitment to accentuate the economic dimension to policy had the 'fingerprints' of British influence all over the legislation.

The decision to adopt the principle of Best Available Technology Not Entailing Excessive Cost (BATNEEC) represented a subtle, but nonetheless crucial, shift away from the principle of precaution which had been mooted in the Environment Action Plan launched with much fanfare during Ireland's 'green' European Presidency. The declaration was greeted with dismay among environmentalists who saw it as further confirmation of the failure of environmental policy to evolve in a more ecologically conscious manner. In Ireland, critics of the EPA Act saw BATNEEC as simply a variant upon its discredited predecessor, the best practicable means (BPM). In Britain, the Royal Commission on Environmental Pollution proposed the Best Practicable Environmental Option to replace 'the best practicable means' which had been the guiding principle of air pollution controls (see Royal Commission, 1976). The problem with BPM was that it left a possible defence from criminal liability if manufacturers could show they were using the best practicable means in their production processes. A considerable period of time elapsed before the British Government eventually responded to the recommendations of the Royal Commission, finally producing the 1990 legislation which enshrined an alternative principle; BATNEEC (Weale, 1996, pp 114–117).

BATNEEC, IPC Licensing and the EPA's Regulatory Ethos

Constructed in a period of eco-political turmoil in Ireland, the IPC licensing regime was designed with the express intention of resurrecting public confidence in environmental regulation. It was, after all, widely acknowledged that the regulatory framework was bureaucratic, unwieldy and had failed to regulate industry. The government was concerned about the lack of expertise among local authorities, the difficulties experienced by developers and industry in obtaining multiple authorisations and the need to establish clear national environmental standards (Scannell, 1995, p. 513). What remains novel in this regime was not so much the political endeavour to improve environmental standards, as the way in which this would be achieved.

In applying for an IPC licence the onus would shift toward the operator to justify and defend the types of technology and practices to be used, a process it was hoped would encourage a new level of environmental awareness among management. Prior to IPC, most firms would have been unfamiliar with either continuous release monitoring, the public disclosure of environmental information or the need to justify publicly its production procedures. As such, it was anticipated that the IPC regime would restrict the capacity for companies to 'cut corners' and simultaneously encourage a new, more holistic managerial approach to environmental protection. In a more innovative fashion, the EPA's objective was not only to ensure a reduction in pollution but that an on-going programme of environmental management and control would focus on *'continuing improvements* aimed at prevention, elimination and *progressive* reduction of emissions would be adopted' (EPA, 1994, p. 2, emphasis added).

The main environmental objective of IPC is to prevent or solve pollution problems rather than transferring them from one medium to another, a situation which tends to create an 'incentive' to release pollution. Although the EPA guidelines suggest that a key aim is to 'eliminate or minimise the risk of harm to the environment' by preventing the emission of potentially polluting substances, it goes beyond the traditional framework of pollution control by encouraging the anticipation of the environmental effect of emissions, not just in the environmental medium into which they are released but also the potential for the emissions to cross over into other environmental media (EPA, 1996, p. 2). This certainly appears a significant step forward and yet the introduction of IPC licensing was not universally welcomed.

In the UK, for example, where it had been introduced as part of the Environmental Protection Act, 1990, there was significant apprehension at the prospect that access to information would allow the 'green nutters to get on parade and have a field day of litigation against industry on entirely inconsequential grounds' (ENDS, 1994, p. 184). Such sentiments have also been echoed in Ireland where property developers expressed dismay at the actions of Lancefort and the decision by the courts to uphold its objections to the venture forwarded by Treasury Holdings (*Irish Times*, 24 June 1998).

However, from management's point of view, there were benefits to be gained. A new, and more focused timescale for (un)successful applications was put in place and, with a single agency, there was the prospect that decisions would be consistent. Above all, business wants to avoid 'unnecessary delay', tying up ventures in a protracted legal wrangle. Better a decision is made quickly, even if it is unsuccessful, so that it can move on.

In granting an IPC licence the agency's decision is underscored by assessing whether the 'best available technology not entailing excessive cost' (BATNEEC) has been achieved. The EPA's guidelines inform us that the technology (which extends to production techniques) should be 'best at preventing pollution, and available in the sense that it is procurable by the operator of the activity concerned'. NEEC would set out the 'balance between environmental benefit and financial cost' (EPA, 1996, p. 2). In order to familiarise business with this new regime the agency published a series of BATNEEC guidance notes which identified the types of technologies which it considered suitable to form a decision on the grounds of BATNEEC. Once established, the agency would then set the relevant emission limit values (ELVs). As its guidelines state, the principle of BATNEEC emphasises pollution prevention techniques, cleaner technologies and waste minimisation, rather then end of pipe treatment (EPA, 1996, p. 3). The notes appear clear cut. Simple. Even intelligible. It certainly appears to represent a quantum leap on the instruments available in the 1980s and yet, on closer inspection, significant difficulties remain. By lending weight to the NEEC element of the principle it is perhaps less stringent than the principle of Best Available Technology (BAT) proposed in the EU draft directive on integrated pollution control (Maguire, 1995, p. 117).

The language which pervades the EPA's guidelines is reassuring, an intoxicating blend of the straightforward and the uncompromisingly incomprehensible; 'state of the art technologies' fuse with the cutting edge of science and its emission limit values. I was tempted to choose the metaphor 'Star Wars meets Trumpton'. However, while most people are familiar with the high-tech special effects of Star Wars fewer people in Ireland are perhaps aware of Trumpton. It was a children's programme on the BBC which contained an enchanting mixture of an idyllic village setting with a reassuring ambience. There's a windmill, a fire station and the usual crop of 'friendly' shop keepers; banker, baker, butcher and candlestick maker (if my memory serves me correctly). There was no gridlock in Trumpton. We can surely brook little argument with the phrase 'state of the art', even if we have no clue about what constitutes an ELV. It is an evasive language, successful because it distracts our attention from tensions which lie at the heart of this principle. Thus, for example, the EPA is emphatic that 'state of the art' should refer to both technology and a 'range of currently employed techniques'. However, in practice it employs an important distinction between 'new' and 'existing' facilities, one which receives rather less attention. As far as new facilities are concerned the BATNEEC equation should have regard to 'the current state of technical knowledge, the requirements of environmental protection and the application of measures which do not entail excessive costs' (EPA, 1996, p. 4). However, the equation alters significantly when we consider existing facilities. Here, an *additional* regard should be had to the nature, extent and effect of the emission concerned, the nature and age of the activity and the period during which the facilities are likely to be used or to continue in operation. Furthermore, a decision to impose conditions on the licence will depend on the 'costs which would be incurred in improving or replacing these existing facilities' and 'in relation to the economic situation of activities of the class concerned' (EPA, 1996, p. 4).

In practice, therefore, we can discern a clear departure from what initially appeared a very tight script. There are no absolute standards or emission levels and, despite the view promulgated by the EPA that 'scientific rigour' underpins the determination of what is an acceptable level of pollution, it is clear that the economic circumstances of an operator have an important bearing upon what emission level is stipulated and how it should be achieved.

The waters are muddied further by the fact that the agency only *envisages* that existing facilities will *progress* towards the attainment of emission limit values similar to those of new facilities (EPA, 1996, p. 4). The distinction between new and existing facilities was also used in setting the dates for the introduction of licensing which produced a number of anomalies. Mineral extraction, for example, was intended to come under the IPC licensing regime by May of 1995, whereas intensive agriculture would come under its jurisdiction only in March of 1996 (EPA, 1996).[12]

That the principle of BATNEEC provides the best direction for environmental policy is clearly a contentious issue. Critics of this policy instrument have argued that what constitutes 'best available technology' or 'excessive' cost is open to considerable debate. Opinion remains strongly divided. In particular, questions have sought to ask whether judgements on these matters measure the external costs of investment or the external costs of not adopting the best available technology. It has certainly provided a fertile ground for debate between the EPA and those licensees who object to more stringent licence conditions on the grounds of cost (*Irish Times*, 28 May 1998).

To environmentalists, one of the more appealing features of the new IPC licensing regime was that it would, for the first time, produce reasonably accurate figures for pollution releases and their environmental impact. It was assumed that such information would perform a dual function, restricting the ability of companies to persist with 'cowboy practices' and simultaneously encourage a new, more holistic managerial approach to environmental protection.[13] In practice, however, these aims have been partially undermined by the EPA's reliance upon self-monitoring and its preferred option to rely on voluntary compliance (Taylor, 1998). The problem with a soft regulatory style such as this is that when it is coupled with a persistent level of non-compliance it undermines confidence in the regulatory regime, serving to reinforce a culture which does not take seriously environmental issues.

As if to compound matters further, it is also a legislative framework which pays little heed to environmental protest. Channels for the improved dissemination of information have been put in place, there is even the possibility that an oral hearing may be conducted, but the general tenor of the Act is upon establishing 'procedural regularity', closing the loopholes exposed by the environmental disputes of the late 1980s. This has been most evident in the procedures adopted for licence applications which were profoundly shaped by regulations adopted in the planning arena. It was legislation which essentially sought to reduce the number of vexatious appeals and curtail excessive delays to major projects. There are, for example, specific obligations imposed on third party objectors (what constitutes a 'valid submission' and the timescale allowed for submission), the information supplied by developers with respect to a valid application (types of technology, production procedures, costs) and, finally, a directive which ensures that the agency makes a decision as 'expeditiously as possible'.

As a consequence the sequence of events, from the publication of the intention to apply for a licence to the decision on whether to grant or refuse a licence, is tightly controlled. If all goes well a licence application can be completed in as little as 2 months, provided that the agency is content with the quality of the application and finds no grounds for refusal after submissions from interested parties. However, if objections are received, then the agency may (or may not) decide to hold an oral hearing, which sets in train a process that may then take up to 4 months to be completed.

The issue of procedural regularity (and the subsequent impact of new legislation on the nature of environmental protest) came to the fore in the controversy surrounding the EPA's decision to allow trials on genetically modified crops. The principal argument in the case brought by Genetic Concern was that the EPA should not have consented to trials for genetically modified crops because it could not ensure that the 'risks to the environment were effectively zero'. Indeed, Genetic Concern were of the opinion that the EPA recognised this when it publicly conceded that the risks involved were only 'very low'. The final judgement, delivered by Mr Justice O'Sullivan, stated that in deciding whether to consent to a deliberate release of a genetically modified organism, the EPA is not required to be satisfied beyond reasonable doubt that the risk of adverse effects has been reduced to an effectively zero level (S Gillane, *Irish*

Times, Law Report, 1998, 14 Dec, p. 26). To sustain this argument he pointed out that the 1992 Act, the 1990 EC Directive and article 33(4) of the 1994 Regulations in relation to the release of GMOs, provides that the agency:

> shall not consent to a deliberate release unless it is satisfied that deliberate releases will not result in adverse effects on human health or the environment (S. Gillane, *Irish Times*, Dec, 14th, 1998, p. 2).

Read together, however, these regulations assume a 'step by step' approach, whereby trials are conducted and evaluated before further measures are taken. In this context, interpreting the word 'avoid' as being 'effectively zero', renders pointless the study of risk and the anticipated further steps in this exercise. In other words, the possible risk of release is inherent to such evaluation, otherwise why bother to prepare for assessments in the first place.[14]

A further issue raised which is of importance to our discussion relates to the nature of a 'valid submission' and the shift in recent legislation (both in planning as well as the EPA Act) to constrain the time period allowed for objections to be constructed. The contention of Genetic Concern was that in a situation where supplemental material had been provided by Monsanto to the EPA in response to a questionnaire, then the third party objector should have been given extra time (more than the 21 days stipulated) to make further submissions. The findings of Justice O'Sullivan on this matter are revealing. He stated that a considerable amount of the opinion evidence would have to be ignored as it was the EPA's function to assess the scientific evidence and the courts to determine the legal issues. It was, moreover, clear that the 1994 regulations did not 'provide for further representations from members of the public after the 21 day period had expired, even where further material had been supplied by the party seeking consent'.[15] The case illustrates two important themes. First, that the legislative framework (and environmental policy in general) is now permeated by scientific jargon, a fact not only recognised by the courts but one with which they are ill at ease. The courts response, reflecting the tone set in legislation, is to rely increasingly upon judgements about procedure. It is this latter issue which remains one of the more prominent themes of the environmental disputes of the 1990s.

Local Authorities and the EPA

If pollution control licensing was the most significant feature of the legislation, it was followed closely by the EPA's assignment to monitor and regulate the role of local authorities. In essence, this part of the Act was intended to redress the widespread problems identified in the role of local authorities in the 1980s. On this issue, the government remained adamant that the EPA would ensure that local authorities face up to their 'environmental obligations' and that it was desirable that a 'mechanism should exist to guide standards of local authority performance and to encourage the consistent application of control procedures around the country' (Harney, *Dáil Debates*, 1991, 411, p. 1261).

It is in sections 56–60 of the Act where the substance of change can be found. Here, the general powers of advice and assistance available to the agency allow it to fix criteria for the management of sewage treatment plants and landfill sites and to perform a general supervisory role in the area of the quality of drinking water. Of more import to the environmental lobby, however, were the powers of sanction the agency now possessed. In cases where the EPA believes that a local authority has failed in its statutory obligations or, where it feels that these responsibilities have been carried out in an unsatisfactory manner, the agency *may* request a report from the local authority. If a local authority fails to respond adequately the EPA *may* implement the necessary changes and impose the costs on the local authority.

At a cursory glance, this would appear to have strengthened significantly the regulatory control of local authorities. However, such optimism is tempered by the fact that the Act is not as robust as many had anticipated. For example, there is no statutory obligation on the part of the EPA to request such reports. As if to confirm further a drift from the intention of bringing local authorities 'to heel', the legislation also allows for convoluted reasons for inaction, a finding confirmed by the observations of one deputy in the Dáil who remarked that, it seems a bit 'naive to expect a local authority to make an objective and useful report on their own shortcomings' (Dukes, *Dáil Debates*, 1991, 411, p. 1711).

Such flaws remain all the more disconcerting when we consider that despite all the 'talk' of a radical departure, and the fact that that several functions previously performed by local authorities have been transferred to the agency, a large number of responsibilities still fall within the ambit of local government. Whatismore, one of the principal weaknesses of the legislative framework of the 1980s, that local authorities were not subject to the same controls as the private sector, remains. As Scannell points out, local authorities are still exempt from having to obtain planning permission for their developments, to obtain a licence for the discharge of sewage and trade effluents to waters or to obtain a permit for the disposal of wastes (Scannell, 1995, p. 517). It is an issue upon which Scannell was moved to remark that while the 'freedom' enjoyed by local authorities under the previous legislation has been eroded, the powers and remedies provided under the EPA Act demonstrate 'the ingenuity of the drafters of the Act in preserving some of the privileges and immunities enjoyed by local and sanitary authorities' (Scannell, 1995, p. 523).

While political debate may have been effused with the emotive language of protecting the environment, a closer scrutiny of the legislative framework reveals that, when presented with an opportunity to restructure a byzantine regulatory framework, government chose not to challenge what lay at the very heart of a discredited previous regime; a soft regulatory ethos underscored by ministerial discretions and opt out clauses. Nowhere is this more evident than in the section 63.3 of the Act which stipulates that the agency cannot issue a directive to a local authority unless it possesses the available funds with which to comply. A remarkable clause by any standards, suggesting that 'if local authorities can be relieved of their obligations because they do not have the funds' then smart local authorities 'will use their funds for everything except the requirements of the Bill and put the onus for its implementation on the Minister' (Dukes, *Dáil Debates*, 1991, 411, p. 1711).

Planning, Environmental Impact Assessment, Exempted Development and IPC

One of the more controversial and confusing aspects of the EPA legislation was the decision not to assign the agency a more prominent role in the planning system, an occurrence all the more perplexing when we consider that the innovative character of the new legislative framework lay in its emphasis upon integration. Planning would continue to operate in the domain of the local authorities and on appeal to An Bord Pleanála. While the EPA's role in this arena has been limited, planning regulations have undergone significant reform which has, in many respects, been beneficial to environmental conservation.

To a large extent, the motivation behind such reform stems from the demands of property developers to change a planning system felt to be excessively protracted and cumbersome. In its appraisal of Irish planning procedure, for example, the Culliton group, which submitted detailed evidence to the Minister prior to the Act, argued that while Irish planning procedures stood up well to international comparison in terms of processing, there were still difficulties with third party objections. The group also made recommendations on the role of the EPA *vis a vis* planning. Its aim was not simply to highlight the need for co-ordination and integration, but to ensure that the EPA legislation dovetailed the timing of its decisions over planning and licensing with An Bord Pleanála (Culliton, 1992, p. 48). The Government's response, the Local Government (Planning and Development) Act of 1992, followed the tendency of legislation during this period to streamline the appeals stage of the development process by focusing on the need to reduce the time period for a decision on appeal to An Bord Pleanála. It was by no means the only change in planning regulations which have had an important bearing upon environmental regulation. Although the EPA's role in this area is limited it is nonetheless invaluable to examine briefly the wider regulatory framework in which the EPA functions and, in particular, the changes which have taken place in the planning arena.[16]

From the outset it is important to recognise that there is a 'presumption of law in favour of development' (Scannell, 1995, p. 187). Indeed, as Scannell points out, planning authorities have been advised by the Department of the Environment that the 'approach to development control should not be unduly restrictive' and that permission should be refused 'only where there are serious objections on important planning grounds' (cited in Scannell, 1995, p. 187). With this in mind this section discusses the most relevant areas in this arrangement which include: the controls and powers available to the planning authority, development plans, ministerial discretions and government policy, exempted development and environmental impact statements (EIS). The legislative pivot in the planning arena is, as Galligan explains, the planning authority, which has two principal functions: development and control (Galligan, 1997, p. 1). The planning authority is essentially the first 'port of call', and ensures that any activity (or change of activity) is carried out in accordance with the development plan. In the process of a planning application it is the body which grants or refuses planning permission. It also has a role in the appeals process and has important powers to enforce and ensure compliance.

The principal instrument for control lies in the zoning of land for use, which is formalised in the development plan. It is, as Galligan notes, a strategic document, designed to ensure orderly development, disseminate information and should contain both long-term and short-term objectives. It has certain mandatory objectives and should be reviewed at least once every five years (Galligan, 1997, p. 21). Once constructed such plans become a crucial means through which developmental control is established. The planning authorities, for example, may constrain development because it is a 'material contravention' of the proposed development plan or a 'material alteration' to existing use. Pressure groups or concerned citizens may also use this particular avenue as a basis from which to construct objections to proposed developments by a planning authority on the grounds that it contravenes its own development plan. There are a number of contentious issues which surround this topic, most notably what constitutes a 'material contravention' of a plan and how the plans are interpreted (see Galligan, 1997, pp 20–37).

This emerged as an issue in the objections raised against Galway Corporation's plans to develop a halting site at Bishop's Field which was not part of its development plan. The Corporation's response was to remove all specific proposals for halting sites from its development plans so that these would no longer be inhibited by their inclusion in the plan. It was a position which goes against the idea, proposed by Mr Justice McCarthy, that the development plan should be perceived as an 'environmental contract'. In this sense, the approach taken by Galway Corporation undermines the 'spirit', if not the letter of the law, behind the development plan and, as Galligan observes, raises the distinct possibility that the terms of the plan 'will become so vague as to give local authorities unlimited powers in deciding where to locate' (Galligan, 1995, p. 1).

A further source of control available to planning authorities lies in the use of Special Amenity Area Orders (SAAO) which enable planning authorities to control land use where planning conditions could not be applied. It is, in a sense, part of a wider package of regulations inspired by EU legislation which gives additional powers to ensure conservation in areas of outstanding beauty or scientific interest. Once established, it allows the planning authority to impose a conservation order for the protection of flora and fauna. One of the more important features of this part of the legislation is that it removes certain types of exempted development, and once designated, protects planning authorities from compensation claims which are liable under 'normal' conditions (Galligan, 1997, p. 39)

The legislative framework also stipulates that planning authorities should have regard to general policy directives issued by the Minister. In instances where the Minister sees fit, he or she *may* require a planning authority to vary a development plan, insist on the co-ordination of plans between local authorities, publish general instructions and issue directives concerning government policy. The use of this latter instruction assumed recent prominence when the government issued the edict that it wanted to curtail the number of large scale shopping centres (Galligan, 1997, p. 10). However, it has also had a bearing on other aspects of environmental conservation where policy at a local level (planning authority) was at odds with general government policy. Here, the legislation dictates that planning authorities and An Bord Pleanála must have regard to any Ministerial directive relevant to the planning exercise. The failure to exercise 'sufficient regard to government policy' was the principal reason behind the failed objections of Mayo County Council to the proposed mining operation by Glencar Explorations PLC, where the courts found that its development plan was 'diametrically opposed' to government policy (Galligan, 1997, p. 29).

The role of wider government policy on environmental conservation and planning issues has also had a significant bearing upon the controversies surrounding the placement of wind farms. Welcomed as panacea to the environmental ills of the fossil fuels, wind farms have recently attracted criticism from certain elements of the environmental lobby. Government policy, however, has indicated a commitment to increase the use of wind farms as a part of its renewable energy programme (Flynn, 1996). This has been influenced by the EU's Fifth European Action Programme on the Environment (Towards Sustainability). It has involved a requirement in which the ESB is committed to purchasing 75MW. The government also announced a renewable Energy Strategy in which there was a target of 30 megawatts of wind generating capacity for each year 1997/8/9. The government has published guidelines on wind farm developments to facilitate planning authorities. The precise legal status of these guidelines is not clear, but in the case of the UK one commentator has observed that:

> The difficulty is that a planning authority, is expected to weigh, on the one hand the advantages of a particular proposal, which will on its own offer only a relatively small contribution to the national objective (the promotion of renewable energy) with, on the other hand, visual and other local impacts which may be only apparent within its locality (cited in Flynn, 1996, p. 144).

In the case *West Coast Farms Ltd v. Secretary of State* in the UK, the judgement was that guidance notes did not create a presumption in favour of renewable energy. Important though these guidance notes were, they should be balanced with policies which protect the environment (Flynn, 1996). Such issues are likely to increase as objections to wind farms materialise.

Environmental Impact Assessment

One of the more noticeable influences of EC legislation in the area of planning, and its effect with regard to environmental protection, has been that of Environmental Impact Assessment. The EC Directive 85/337/EEC came into force in 1988, although it was not introduced into Irish legislation until the Local Government (Planning and Development) Act, 1992 and the Local Government (Planning and Development) Regulations 1994 were enacted. Once again, it is an area which has attracted considerable controversy, primarily over the issue of when an EIS is required and when it is not. In part, this can be attributed to the fact that a directive does not become part of Irish law automatically and that EC Regulations allow significant discretion to Member States to interpret what should require an EIS (Scannell, 1995, p. 289). Difficulties in this area abound, and, more often than not, are associated with the lack of confidence people have in defining what an 'integrated chemical facility or an urban development project is' (Scannell, 1995, p. 290).

The Environmental Impact Assessment (EIA) regulations contain a series of clauses which stipulate the nature of information to be contained in an environmental impact statement. They include, among other things, reference to the environmental effects of the project on fauna and flora, soil, water, air, climate and the landscape. As with other areas of Irish environmental regulation, the decision to introduce new legislation was avoided, which has meant that 'grey areas' persist. This, as Scannell notes, was probably a result of 'resistance from the public sector developers who have, until recently, been exempted' from the control imposed on the private sector (Scannell, 1995, p. 293).

On a more upbeat note, the EPA Act goes further than EC requirements on the dissemination of EIS information, since it requires not only that a notification of an exemption be made, but that the agency be consulted for its views which are then made public (Scannell, 1995, p. 290). The Environmental Research Unit also found that there had been considerable improvement in the quality of EIS submissions in the late 1980s and, as Scannell observes, there has been welcome enthusiasm among Irish authorities to insist upon EISs.[17] Unlike its counterpart in the UK, for example, all extractive industries and pharmaceutical and chemical companies require an EIS (Scannell, 1995, p. 311). The exception to this trend of improvement lies in afforestation and land drainage, areas in which the Irish State has a significant vested interest in expansion (Scannell, 1995; Culliton, 1992).

Planning, IPC and Masonite

Planning authorities and An Bord Pleanála have traditionally been involved in environmental protection in the assessment of land use. Prior to 1994, issues of environmental pollution associated with proposed or existing developments were assessed by local authorities and controlled through conditions attached to planning permission. Appeals were taken to An Bord Pleanála. However, section 98 of the EPA Act has shifted the terms of responsibilities so that where an IPC licence is required, the EPA is the competent authority when assessing an EIS, but only insofar as it relates to environmental pollution matters; all other considerations (landscape, visual effects, traffic implications etc.) remain within the ambit of the planning authority and An Bord Pleanála.

This division of responsibilities between the EPA and the planning authorities on projects which require an IPC licence has caused considerable confusion. It is a complicated issue, but one which is important. Put simply, the effect is that neither the planning authority or An Bord Pleanála can consider matters relating to *potential* environmental pollution from a project where an IPC licence is required. Pollution generated during the development phase may be considered, but not pollution which is likely to occur once the plant is up and running. This can have significant repercussions for the respective roles of the EPA, the planning authorities and An Bord Pleanála and came to light most prominently in the Inspector's report on Masonite.

The proposal by Masonite to construct a fibre manufacturing plant in Leitrim had been objected to by numerous parties who argued that, while issues of water and air pollution may not be considered by the Board (following section 98, EPA Act) they felt that such matters were subjective perceptions regarding pollution risk and, therefore, were relevant planning matters (Brassil, 1996, p. 21). However, the Inspector concluded that 'perception' was a matter relating to the risk of environmental pollution (once the plant was up and running), a factor which could be considered only by the EPA.

In a further issue raised in this case, Masonite suggested that the Board and the EPA must take a balanced view of the proposed development with regard to 'environmental protection and infrastructural, economic and social progress' (Brassil, 1996, p. 22). The Inspector's response was to point out that 'the sub-division of responsibilities resulted in 'two separate balancing acts, in which the benefits of the development are weighed against separate sets of drawbacks or disadvantages' (Galligan, 1996, p. 2). Given that the two bodies may utilise different criteria in terms of costs and benefits it was quite possible they may reach 'two conflicting recommendations' (Brassil, 1996, p. 22).

In part, this sub-division of responsibilities has clearly taken some of the 'teeth' out of An Bord Pleanála, restricting its capacity to impose planning conditions. However, it is also worth pointing out that in the event that the planning authority or An Bord Pleanála find in favour of a particular development, it is likely to increase the 'pressure' upon the EPA to recommend the proposal, no doubt with conditions, but recommend nevertheless. Government is unlikely to be too enamoured if a major project is accepted by An Bord Pleanála and then finds that it is turned down by the EPA. After all, its function essentially is to engineer a situation in which compromises can be reached, to ensure that firms *become* compliant, and not to prevent development.

It is commonplace in cases such as this to pursue the possibility (where objections arise) for alternative sites to be considered, an avenue which was taken by third party objectors. However, it was an instance in which the sub-division of responsibilities between the Board and the EPA raised further issues for concern. Masonite had argued that an adequate body of water for the disposal of effluent was necessary in its site selection, a point on which the Inspector concurred, noting that once again these considerations no longer remained in the ambit of An Bord Pleanála and were the responsibility of the EPA. It has undoubtedly been a source of consternation among the planning authorities since it effectively precludes:

> the Board from consideration of the entire range of site selection criteria, thereby obviating the possibility of a comprehensive and fully informed assessment by the authority ultimately charged with locational considerations (Brassil, 1996, p. 23).

In a more cautionary tone, the report concluded that the sub-division of the appeals process between the EPA and An Bord Pleanála 'may not be as comprehensive or as effective as the concurrent consideration of land use and environmental pollution matters by the local authority and An Bord Pleanála' (Brassil, 1996, p. 23).

Exempted Development and the Mullaghmore Controversy

The issue of exempted development (development which does not require planning permission) was a crucial factor in the environmental disputes which took place over the proposed interpretative centres to be located at Luggala and

Mullaghmore. In general terms, exempted development (which had been thought to extend to state authorities) was justified either on the grounds that a reduction in administration costs could be achieved with minimal environmental harm or that development undertaken by the state was in the interests of the common good (Galligan, 1995, p. 84). In short, the Mullaghmore case set out to question whether the visitor's centre, which was already under construction, was exempted development. The issues which arose in this case revolved initially around section 84 of the 1963 Local Government (Planning and Development) Act, which prescribed a procedure:

> whereby state authorities were *obliged to consult* with the planning authority in relation to construction or extension of any building, and if objections raised by the planning authority remained unresolved, they were obliged to consult with the Minister (Galligan, 1997, p. 100, emphasis added).

The objectors to the development argued that a necessity to *consult* would surely not have arisen if it was intended that the state should be exempt from planning permission. On appeal to the Supreme Court they continued with this line of thought, contending that the purpose of the *obligation to consult* was to avoid the cost and expense of an *'unsuccessful planning application'* (thereby linking the obligation to consult with a planning application and the need to have permission). Moreover, it was in their opinion also undesirable for state agencies to be engaged in projects which were inconsistent with the policies of a planning authority. Opinions in the Supreme Court were divided, but the final judgement, delivered by J. Blayney, found that the state did not enjoy any general exemption from the obligation to obtain planning permission under section 24 of the 1963 Act and, given that the interpretative centres also involved the construction of structures which were not buildings (car parks, waste treatment etc.), they required planning permission.[18]

Finance, Independence, Accountability and the EPA

During the passage of any legislation through the Dáil the opposition parties often propose objections, delay procedure or seek grounds for amendment. If we ignore those contributions devalued by the tendency to 'score' political points, important issues are often raised. In this sense, the legislative passage of the EPA was no exception. In particular, a number of prominent themes attracted the attention of the opposition; the independence of the agency, finance and the problem of accountability, all of which have assumed greater prominence since the inception of the agency.

During the embryonic stages in the formation of the EPA bill, the Minister of the Environment and her colleagues contended that the agency would possess both the requisite regulatory powers to protect the environment and were adamant that it would remain firmly independent of government. When drafting the legislation, however, there was a discernible drift in the government's position on both counts. The failure to make definitive commitments on funding, the lack of an appeals structure for environmentalists and the very composition of the first advisory committee to the agency all revealed a greater concern with the media's reception of the agency than with the minutiae of its administrative and institutional aims (Taylor, 1998).

The composition of the first advisory committee was the first area to provoke criticism from the environmental lobby. The function of this committee was to assemble a list of candidates from which the Director General and four directors of the agency would be selected. The council has a three year life span after which time its members change. However, the Act stipulates that a number of bodies should have permanent representation; the IDA, the Irish Congress of Trade Unions (ICTU), the Council for the Status of Women and An Taisce. The environmental lobby were scathing in their criticism of IDA's representation on the advisory committee, arguing that it was almost an incomprehensible decision to have an organisation like the IDA on the Board of an independent environmental body when they are pro-development, pro-industry, and represent everything that is anathema to environmental policy (Taylor, 1998).

The Minister of the Environment, Mary Harney, remained unperturbed by such criticism, defending the inclusion of the IDA on the grounds that if an inappropriate company located in Ireland then it was the legislation and controls which were at fault, and not the 'body charged with the task of aggressively selling Ireland and promoting industrial development' (Harney, *Dáil Debates*, 1992, 418, p. 959) It was, in her considered opinion, more apposite than an arbitrary decision between the Confederation of Irish Industry, the Federation of Irish Chemical Industries and the Federation of Irish Employers (Harney, *Dáil Debates*, 1992, 418, p. 959).

Controversial though this issue may have been, the Minister's most pressing concern lay in the need to avert the possibility that the agency would be misconstrued as a baton with which to 'club' multi-nationals into submission. While the Minister of State had been a party to expanding the EPA's remit, she was particularly anxious to avoid the possibility that the EPA would be seen as anti-industry and, while it would regulate industry and other activities with polluting potential, it would not strangle economic enterprise (Harney, *Dáil Debates*, 1991, 411, p. 1263). The Government's position was unequivocal and an issue upon which there could be no misunderstanding; it was not going to be a party to picking a 'group of environmentalists' to review licences (Harney, *Dáil Debates*, 1992, 418, p. 959).

This view was reinforced in the government's stance on the composition of the advisory committee. From the outset, the government was intent to instil a managerial ethos at the helm of the agency and while this would not *necessarily* impose any undue constraints upon its environmental priorities, especially its scientific and research output, it would ensure that the agency's primary focus would be upon *managing* rather than *conserving* the environment (Taylor, 1998). The agency would monitor, assess and, as a last resort, consider punitive action. Above all else, its function was to act as a watchdog, not a bloodhound.

The debate on the advisory committee also formed a focal point for extended criticism of the level of independence the agency would enjoy. As with other parts of the Act, section 21 affords the Minister a number of discretions, which in this case pertain to the list of bodies eligible to submit candidates. The final decision on who is selected rests ultimately with the Minister. As far as the opposition benches were concerned, it raised the prospect that such control may impose a 'conservative outlook' on the agency, compromising its commitment to survey politically contentious areas.[19] On a more conjectural note, Deputy Gilmore asserted that An Foras Forbartha, the precursor to the EPA, had 'come a cropper' because government 'had decided it did not want an independent agency to deal with environmental issues' (Gilmore, *Dáil Debates*, 1991, 411, p. 1297, Taylor, 1998).

A further issue raised in the political exchanges over the agency's independence relates to the thorny issue of finance. During the early stages of the passage of the legislation the government estimated that the agency would need around £8 million. However, its funding has operated at nowhere near this level. In 1994, the agency's funding stood at £4.4 million (Browne, *Dáil Debates*, 23/6/1994, p. 573). In the following year the Minister declared that its available funds would be £3.6 million for current spending, £750,000 capital spending, and an expected fee income of £1.4 million (Howlin, *Dáil Debates*, 27/4/1995, p. 337). It was a figure which fell considerably short of the original £8 million and, on closer inspection, it became quite clear that despite the political rhetoric of its embrace of a new relationship between the environment and the economy the Government was unwilling to release additional funds. Indeed, the funding of the agency, has been steadily eroded and transferred from other areas or raised by the agency itself.

There are few who would argue that such independence does not have its advantages. However, the flipside to this political coin is that the more independence the agency secures the stronger the likelihood that it compromises the capacity of the Oireachtas, a publicly elected body, to scrutinise its actions and ensure accountability. It was a matter upon which even the government's most vociferous critics voiced concern, pointing to the possibility that the agency could become a committee of experts devoid of any channel through which democratic accountability and public participation could be ensured. It was a possible scenario complicated further by the fact that many of the provisions contained in the Act would be operationalised through Ministerial action and, therefore, not subject to debate in the Oireachtas. Opposition members were also perturbed by the prospect that the agency would be separated from the responsibility of the Minister, curtailing the opportunity for public scrutiny. Neither does the Act allow public access to the report of the EPA on an oral hearing. In addition, Section 108 of the Act, allows the Minister to make regulations with regard to access to environmental information and has allowed the Minister to delay the introduction of the relevant EC Directive. Thus, with regard to large portions of the Act, once passed, it is removed from the public arena.

One could be forgiven for thinking that the Government's critics wanted their 'bread buttered on both sides'. On the one hand, they were calling for an independent agency and yet, on the other hand, they were demanding that it should be made more accountable to the Dáil. This apparent contradiction reflects a tension which resides at the very heart of the democratic parliamentary process. It is a theme to which we will return later.

Concluding Remarks

It would be difficult to sustain a coherent argument which suggested that important strides had not been made in the legislative framework that has evolved in the 1990s. The commitment from within the EPA to provide access to environmental information and the steps undertaken to introduce new and innovative monitoring techniques are welcome initiatives. Moreover, the willingness on the part of planning authorities to insist on environmental impact statements and the subsequent improvements in the quality of those submissions also give grounds for optimism. However, this should not detract from the significant problems which remain.

In many ways, the EPA's favourable reception stands as a worthy testament to an era in which political spin has become the norm. Headlines may grab the attention, but more often than not at the expense of substance. The legislation undeniably contains clauses which, if invoked, could lead to the director of a company being jailed for 'persistent' pollution offences. There is, however, a vast difference between legislative intent and regulatory practice. It should surprise few that, in a society where amnesties for financial irregularities abound, and where political luminaries have outstanding tax liabilities which exceed the lifetime income of the average citizen quashed, scepticism of the agency's commitment to enforce the regulations persist. Whatismore, beneath the facade of radicalism, significant flaws endure, nowhere more evident than in IPC licensing and BATNEEC, which reveal rather more in the way of continuity than change with the discredited regulatory framework of the 1970s and 1980s. It is to this latter issue that we now turn.

Custodian, Guardian or Absentee Landlord?
Environmental Management and the EPA in the 1990s

Introduction

As APW Malcomson, the historian suggests, politically charged though the term 'absentee landlord' is:

> absentees ... were not unconscious sinners – if indeed they sinned at all. On the contrary, they were an acutely self conscious body of men; and if in some this self consciousness expressed itself most obviously in sensitivity to threats to their vested interests, in others it expressed itself laudably in a sense of special obligation (Malcomson, 1974, p 15).[20]

It is a term 'bedevilled by the assumption that absenteeism implies improvidence and neglect', an allusion toward the landlord's (mis)conduct and how estates were 'managed or rather mismanaged' (Malcomson, 1974). And yet, it is also a term which conveys the view that, in the absence of a landlord a moral or political vacuum is left, a pervading sense of ambivalence in which there is no one to lead by example. The title therefore seeks to capture a tension which resides both in the term absentee landlord and the regulatory role of the EPA.

For many of the EPA's critics, there is an intense disillusionment at the failure of the agency to 'live up' to the weight of expectations which surrounded its arrival. It has not, in their opinion, led by example, acting more as a *watchdog*, rather than the anticipated *bloodhound*. The institutions may have changed and new regulatory responsibilities taken on board, but its critics have failed to realise that it is the underlying regulatory ethos which remains intact, and it is this, rather than any particular institutional format, which is the source of angst. There is also a further area of concern, one which surfaces constantly throughout this chapter but is largely misunderstood; that is the extent to which the EPA's regulatory responsibilities remain limited. As we shall see below, in many instances the EPA is not always the competent regulatory authority; its responsibilities are often sub-divided and significant Ministerial discretions persist. The consequence has been that while it was touted originally as a new super agency, with powers vested in it to perform a myriad of tasks, the original optimism has given way to cynicism. The chapter seeks to explore both the EPA's role and the changing nature of environmental regulation during the 1990s.

Agri-Environmental Policy and the EPA

The EPA has both direct and indirect responsibilities for regulating pollution arising from agriculture. It is directly responsible for licensing large scale pig and poultry operations and, once licensed, operators are expected to have in place a self-monitoring plan that is checked on a random basis by the EPA (Sherwood, 1994, p. 62). Few would disagree that the regulation of pig and poultry farming was in urgent need of reform, a process in which it was anticipated the EPA would play a significant role. Indeed, in one of its earlier sets of forecasts, the EPA suggested that new pig and poultry operations would be licensable by the end of 1994, and that existing operations would be brought into the system within twelve months (Sherwood, 1994, p. 62). It was a forecast which proved to be rather optimistic, probably because the agency had underestimated the trenchant and well organised opposition of the farming lobby. The introduction of IPC licences for pig and poultry units emerged only in September 1996 and licensing for existing operations did not commence for a further year.

From the outset, the agricultural community was openly hostile to any licensing scheme and in order to *reassure* the Irish Farmers Association (IFA) the Minister for the Environment acceded to requests that the licences would be open-ended, with no need for renewal after a specified period (*Farmers Journal*, 10 August 1996). The IFA was also upset at the perceived intransigence of government in failing to negotiate change with the industry. In a move designed to remove any repetition along these lines, the IFA decided to bring the issue of licences (and their attendant fees) within the remit of the *Programme for Competitiveness and Work* (PCW). As a consequence, the IFA was successful in delaying the extension of the 1997 Ministerial Order to existing pig farm operations until March 1998.

The EPA is also involved in regulating the environmental consequences of agricultural activity in the area of water pollution, where the excessive use of fertiliser or slurry leads to nutrient run-off and contributes to eutrophication. With the assistance of local authorities, the EPA undertakes monitoring and, in instances where the quality of water is found to be deteriorating, the agency is expected to take action to rectify the situation. However, all local incidents of water pollution continue to be the responsibility of local authorities, the Fisheries Boards, and affected parties who can prosecute under existing legislation. As a result, the arrival of the EPA has not precipitated a notable change in the regulatory regime at this level.

As in other areas of its responsibility, the EPA's preferred approach is to avoid stringent environmental controls reinforced by prosecution. While options to reduce the excessive use of fertilisers and the spread of slurry exist, the EPA prefers to coax farmers into adopting new practices and has argued that:

> it is far more important that farmers should become fully aware of all the circumstances under which nutrients from agriculture reach water and that they should *voluntarily adopt practices to prevent it* ... it may take time but the prize is worth working for (cited in Sherwood, 1994, p. 66; emphasis added).

Thus, despite the fact that the EPA's report on the quality of water in Ireland 1991–1994 clearly identifies the agricultural sector as the single largest cause of water pollution, it has sought to reassure the agricultural community that the agency does not pose a significant threat to the status quo. It is a position upon which the agency has been unequivocal, keen to insist that its messages should not be taken as a threat to agricultural interests, but as early 'warning signals', designed to allow minor changes in policy at an early stage and prevent serious, even irreversible environmental damage at a later stage (EPA, 1996b).

It was a theme which Brendan Howlin, Minister for the Environment, warmed to when he stated that, despite an investment programme of £1.2 billion up to the year 2005 to minimise pollution:

> we are still depending on the goodwill and vigilance on the part of every user of the natural environment, *particularly farmers*, to ensure that accidents ... do not recur (Howlin, *Dáil Debates*, 1995, vol. 453, p. 1981; emphasis added).

In an effort to demonstrate the government's willingness to tackle the threat to water from agricultural waste, the Waste Management Act, 1996 includes an amendment to the Local Government (Water Pollution) Act, 1990 which provides local authorities with *discretionary* powers to oblige farmers to submit a nutrient management plan.

The government rejected arguments from the opposition benches that the discretionary element to the legislation should be dropped, referring to the EPA's report *The State of the Environment in Ireland* which concluded that the use of chemical fertilisers could be reduced by an amount worth £25 million per annum without adverse effects on production. The Minister was optimistic of the possibility of achieving 'managed change' and convinced that, with the combined incentives of the 'good defence' clause and the financial benefits accruing from a nutrient management plan, all farmers would voluntarily comply.

As a consequence, the Waste Management Act exempts from IPC licensing control 'sludge from a facility operated by a local authority for the treatment of water or waste water, blood of animal or poultry origin'. When questioned on the issue of why farm sludge was not included in the Bill, the Minister replied that the 'Bill is not meant to be a comprehensive one' and that 'farm activity and its potential to cause pollution is covered under the licensing regime of the Environmental Protection Agency Act' (Howlin, *Dáil Debates*, 1996, vol. 460, p. 11). It certainly raises the query that if a farmer is compliant within the conditions set out by an 'effective' nutrient management plan, how serious would the pollution have to be to warrant punitive action from a local authority? Of more concern to the environmental lobby was the likelihood that DAFF and Teagasc would strongly influence the specifications for nutrient management plans and, therefore, the power to define the parameters of acceptable farm practice would remain firmly within the grasp of the agricultural policy community. There are at least three other areas of agri-environmental policy which deserve our attention: the Rural Environmental Protection Scheme (REPS), the National Heritage Areas (NHAs) and EU Directives.

The Rural Environmental Protection Scheme is a critical element of Ireland's agri-environmental policy and forms part of a wider package of reform which operates in tandem with the Common Agricultural Policy (CAP). Pre-reform CAP, with its price control mechanisms, had fulfilled a key objective of the Treaty of Rome; providing adequate food supplies at reasonable prices. However, policy was constructed largely around incentives to increase production which had a deleterious effect upon the environment. Both wildlife habitats and water quality were impaired as intensive agriculture increased its use of fertilisers and pesticides (Taylor, 1998). A further unanticipated consequence of CAP was that huge quantities of surplus produce would have to be stored at the tax payers expense. It was a system which had, by the late 1980s, generated a formidable level of opposition as a range of political groups demanded reform (Regan, 1994).

From within the agricultural lobby, suggestions for reform attempted to reconstruct the farmer as both producer and custodian of the countryside (Regan, 1994, p. 75). It is the simplicity of this vision which has undoubtedly proven crucial to its widespread reception – tapping into the image of 'small farmers' tending their flock/environment. It certainly struck a chord with at least one Minister for whom it was incomprehensible that anyone could suggest that farmers were not the custodians of our natural environment because 'they are so directly affected by it, they are more attuned to its protection than most people' (Howlin, *Dáil Debates*, 1996, vol. 465, p. 370). It is a picture taken unmistakably in soft focus. It has, for example, long been recognised that agriculture is an industry with massive polluting potential (see Dodd and Champ, 1983). To a large extent the Rural Environmental Protection Scheme was the policy expression of this duality in the farming function, serving to reconcile the need to maintain income transfers to farmers and yet assuage the environmental lobby's demands for greater regulation of agriculture. It was intended that under the auspices of an environmental conservation programme, government would be able to effect a reduction in the burden of agricultural support, a move made all the more meaningful when placed against the backdrop of a political climate in Europe which was increasingly hostile to agricultural subsidies (Taylor, 1999).

As a policy REPS had three principal objectives. First, to establish farming practices which reflect a concern for conservation, landscape protection and wider environmental problems. Second, to protect wildlife habitats and endangered species of flora and fauna. Finally, it aims to promote quality food production in an environmentally friendly way. Comprehensive, integrated, even laudable.

At its pivot is the nutrient management plan (NMP) which aims to ensure that nutrients are used in a more effective, environmentally conscious manner. It also strives to encourage new measures for the collection, management and disposal of farm wastes. However, there are flaws within its design which militate against its success as an effective instrument for environmental protection. In particular, payments are made only up to 40 hectares which acts as a disincentive for the larger, more intensive farmers. Consequently, there are high participation rates among cattle and sheep farmers – and among farms in the north and west – but it is less relevant to more intensive farming – south west and south east (Leavy, 1997). Moreover, as a survey of farm advisors by Teagasc illustrates, for many of the intensive farms the eligibility conditions were considered too onerous. While a significant minority of farm advisors (25%) felt that participation in this scheme would be environmentally beneficial, there was a perceived difficulty in creating an incentive to encourage participation. And yet, perversely, it is these operations which need to embrace new agri-environmental practices.

A further flaw in REPS is that while aid is available for additional conservation measures, payment for only one measure is allowed. There is little incentive, therefore, for farmers to develop a range of environmentally sound working practices. This has been exacerbated by the fact that farm advisors (69%) felt that there was a perceived conflict between 'efficiency and environmental objectives' (Leavy, 1997, p. 13). If the findings of the Teagasc survey reflect accurately the national picture, the prognosis for future improvement appears rather pessimistic, since the survey found that only 24% of farm advisors thought that 'advice emphasising technical efficiency will help to remove this conflict' (Leavy, 1997, p. 13).

From its inception REPS was beset with problems. Initially, DAFF had predicted a participation level of about 40,000 farmers, covering about 1.3 million hectares. Yet, by the close of 1996, agricultural consultants had declared that participation would have to double for the scheme to reach this target. Such a low level of participation undermined the potential success of the scheme, since a positive contribution from REPS could be realised only if there was a good response to the scheme among qualifying farmers. The survey undertaken by Teagasc suggested that the low levels of participation could be attributed, in part, to the capital cost of the work involved, a feature that 41% of advisors felt militated against participation.

The inspection level of participants in REPS also cast doubt on its credibility as a bona fide pollution control programme. Originally, the intention had been to check all cases for compliance, a move the Minister for Agriculture, Food and Forestry thought was in the best interests of Irish farmers, and that the 'enhancement of the agri-environment' could be served only by 'operating effective control and monitoring arrangements' (Yates, *Dáil Debates*, 1996, vol. 467, p. 1870). Opposing what it saw as unnecessarily stringent regulation, the IFA called on the Minister to halve the rate of inspection in order to speed up the payments to farmers (Taylor, 2000). It was a 'request' to which the Minister duly complied, suggesting in defence of his 'U' turn that most farmers were complying with their obligations under the scheme (*Irish Times*, 26 November 1996).

The credibility of REPS as part of an integrated environmental protection package was undermined further by the fact that it sent out conflicting signals to farmers since, as part of CAP, farmers were encouraged to maintain or intensify farming operations by price supports and annual payments. The Ewe Premium Scheme in Ireland is typical of this initiative. It is a payment for the unit of livestock rather than by hectare, which makes overgrazing financially beneficial to certain farmers.

A further problem facing REPS is that it is largely viewed simply as one of many competing programmes of income transfer, rather than as an environmental protection scheme. This is confirmed by the Teagasc survey, which found that the annual REPS payments form '50% of total farm income in 40% of farm participants' (Leavy, 1997). With a finger firmly on the electoral pulse, it was an issue with which the Minister for Agriculture, Food and Forestry was fully cognisant, observing that REPS was a 'real money spinner for farm families' (Yates, *Dáil Debates*, 1996, 462, p. 1046).

As an environmental protection scheme REPS also has an interface with the programme for the protection of Natural Heritage Areas (NHAs), a programme also beset with difficulties. To a large extent, the obstacles surrounding the implementation of the NHAs have their origins in the European Directive on the Conservation of Wild Birds (Birds Directive) and the legal challenge to An Foras Forbatha's designation of Areas of Scientific Interest (ASI).

The Birds Directive emerged from concern at the EU level with the migratory patterns of birds and the need to protect their habitats (later termed Special Protection Areas, SPA).[21] The principal source of difficulty lay in the politically contentious issue of designating specific sites, a move anticipated to raise the ire of landowners because it could have a negative impact on land values. Between 1995 and 1996, eight set of amended regulations were in place and 106 SPAs were designated. Land ownership largely dictated the boundaries with a preponderance of state-owned areas. The designation of SPAs followed closely the ASIs established in a survey undertaken by An Foras Forbatha which identified the botanical, zoological, ornithological or geological value of a particular site (Grist, 1997).

Since the 1970s there has been very little undertaken in the way of monitoring these sites. It should come as little surprise therefore that a survey in 1983 found that of 80 sites in Donegal, 25% were damaged or threatened. It was a situation which had deteriorated markedly by 1992 where 37% of sites were damaged and a further 16% were under immediate threat (Hickie, 1997, p. 15).

From the 1980s onwards, local authorities have included ASIs in their development plans and yet, an analysis of planning decisions on ASI sites confirms that they were not considered as 'no go' areas. Indeed, in the case of Cork County Council its development plan explicitly stated that:

> listing in the development plan is not always sufficient to protect areas from development pressures, and land of high scenic amenity or ecological value but low agricultural value is particularly under threat (Cited in Hickie, 1997, p. 18).

The designation of ASI status emerged as a significant issue in the application to build an airport on the Errisberg/Roundstone Bog near Clifden, where planning permission was refused, on the grounds that it would disturb the balance of a delicate eco-system and was contrary to the proper planning and development of the area (see Grist, 1997). However, the landowners successfully sought relief against the designation of the land because they had not been notified of the ASI status. As a consequence all designated sites had to be cancelled. It is a politically contentious issue, one which raises the spectre of conflict between EU Directives and constitutional property rights, a problem which figured prominently in the passage of the EU Habitats Directive.

The Department of Arts, Culture and the Gaeltacht (DACG) is responsible for the implementation of the 1992 EU Habitats Directive which forms the 'second pillar' in the European Union's wildlife protection policy and aims to establish a network of protected ecological sites across Europe. While the selection of sites was to be made at the national level, the final decision would ultimately rest with the EU and, in exceptional circumstances, it may add to the list provided by Member States. The Directive makes a provision for a three stage procedure leading to the creation of the Natura 2000 network of sites (Special Areas of Conservation). Stage One, the completion of national lists, should have been achieved by June, 1995 but none had materialised by March, 1997. Some countries had submitted a substantial if incomplete list, but four countries (which included Ireland) had not submitted at all. The others were Germany, France, and Luxembourg. Stage Two of the process (1995-1998) should have established the list of Sites of Community Importance (SCI's). The final stages (1998-2004) should see the formal adoption of the list of SCIs (Grist, 1997).

Innovative and ambitious, it was a landmark attempt to intervene in the control of land use in Member States. Negotiations with Member States, however, proved convoluted and tortuous. Ireland was no exception. The Directive, which should have been implemented in 1994 was still under consideration in 1996. Amid uproar in the Dáil, the Minister for DACG, M.D. Higgins, attempted to impress upon the opposition benches that further vacillation on this legislation would be unacceptable to the EU Commission. Under the legislation landowners are informed of proposals to include their property in a SAC and given three months to object on scientific grounds. One of the thorny issues which arose, and attracted considerable attention in protracted negotiations with the IFA, was that the Directive should not disregard Constitutional property rights. Regulation 20 therefore enables the Minister to make payments through the REPS scheme (Grist, 1997, p. 93).

Delay had been due largely to the strength of opposition mobilised by the agricultural lobby, incensed by the failure to broker a compromise with the Minister. The Minister's response was to indicate that at 'all times' his Department had adopted an approach based upon the three C's: consultation, compromise and compensation. And yet, it is clear that not all interest groups within the Irish political milieu are able to secure the level of access to policy granted to the agricultural lobby. The Irish Peatland Conservation Council, (IPCC) for example, was granted one meeting with DACG before Christmas 1996, at which they were simply shown a draft copy of the regulations and told that they could not be changed.

Negotiations with farmers, on the other hand, continued until mid-February, when the farming press reported that the agricultural lobby had succeeded in securing a major breakthrough on the issue of compensation negotiations. Two packages were agreed: one for REPS participants in SACs, and one for farmers in SACs not in REPS. For those in REPS, the 40 hectare ceiling on payments was raised. Those not in REPS would follow a special plan developed by the DACG and would be required to follow the details of agreements already reached with agricultural organisations on NHAs (*The Farmers Journal*, 22 February 1997).

The latter part of the analysis of REPS illustrates how an agri-environmental scheme interacts with other policies in such a way that it is difficult to substantiate the environmental dimension. Reaffirming the tension which resides in the need to reconcile environmental protection with income transfers to farmers, the Minister declared that apart from protecting habitats, 'it would put much needed income into the pockets of smallholders' (*The Irish Times*, 27 February 1997). This failing has been compounded by the conspicuous lack of a fully *integrated* approach to environmental policy, most evident in the squabbles which broke out between different departments over the controls on afforestation and the debacle surrounding the collapse of the Control of Farmyard Pollution Scheme (CFYPS).

The office for the Minister of State at DAFF declared to the Dáil in January, 1995 that environmental considerations would not be forgotten in the state's afforestation programme and that a review of planning controls on large scale forestry projects would be carried out in co-operation with the DoE (Deenihan, *Dáil Debates*, 26 January 1995, p. 539). The environmental dimension to this policy focused largely on changes to planning permission, where new controls would reduce the threshold for which planning permission and EIAs are required from 200 hectares to 70 hectares. While the reduction in the threshold at which planning permission and EIAs are required was a welcome step in environmental control, difficulties with regulation are still evident, particularly given the poor record of local authorities in the environmental arena.

More disconcerting was the fact that the forestry sector is riven with ambiguities, displayed by the low level of integration between Government departments. It is, for example, by no means clear what the respective regulatory responsibilities of DAFF and DoE are in this area (Taylor, 2001). On this matter, the Minister for the Environment informed the Dáil that the decision on the type of trees planted was primarily a matter for DAFF, while the environmental impact and planning requirements were in the jurisdiction of DoE (Howlin, *Dáil Debates*, 1996, vol. 465, p. 458). In response, the Minister of State at DAFF, when asked who would be responsible for environmental impact and planning controls of forestry, declared that the Forestry Department would be responsible for *all matters* effecting the forestry programme. When it was pointed out that the Minister of the Environment had seen certain aspects of the forestry programme falling within the remit of the DoE, the Minister repeated that 'our Department will be responsible for all matters that we deem affect the implementation of our programme' (Deenihan, *Dáil Debates*, 1995, vol. 454, p. 1334). Such a failure to achieve co-ordination between Departments should come as little surprise to those conversant with Irish politics, where political prestige is lauded upon those who successfully secure or defend the territory of their respective Departments.

It had been intended that the formation of the EPA would reduce such deficiencies in policy implementation. Indeed, the Minister for the Environment had gone as far as to declare that 'all state agencies will dovetail to ensure that we strive for the pristine environment which is our common objective' (Howlin, *Dáil Debates*, 1995, vol. 452, p. 337). It is also a matter upon which the EPA Act expressly states that the agency:

> may of its own volition, and shall, when requested by a Minister of Government, give advice or make representations for the purposes of environmental protection to any such Minister, on any matter relating to his functions or responsibilities (EPA Act, 1992, p. 40).

And yet, the candid remarks of one official at DAFF reveal that this element of the EPA's remit had certainly not seeped through the labyrinthine structures of the Irish civil service sufficiently to have been brought to his attention:

> my Department has to date not found it necessary to consult the Environmental Protection Agency in relation to our forestry programme, nor has the Environmental Protection Agency found it necessary to approach my Department requesting information or offering advice and/or recommendations on our forestry policy and practices (Deenihan, *Dáil Debates*, 1995, vol. 454, p. 1332).

Thus far, any positive change the EPA could have made has been hampered by a powerful agricultural lobby and the unwillingness of individual Government departments to embrace the 'fanciful notion' of integration. It is an argument endorsed fully by the reaction of the DoE to the collapse of Control of Farmyard Pollution Scheme (CFYPS) which attests further to a failure to generate any genuine spirit of institutional co-operation on environmental policy.

The CFYPS was a central part of the government's effort to address agricultural pollution and its subsequent collapse had a deleterious effect on REPS. When questioned about the collapse of the CFYPS the Minister for the Environment claimed that questions in this regard should be addressed to DAFF, and that they had 'no implications for my Department' (Howlin, *Dáil Debates*, 1995, vol. 453, pp 2000–2022). The position held by DAFF was markedly different. As far as the Minister for Agriculture, Food and Forestry was concerned, the collapse of the CFYPS would impact upon the ability of farmers to reduce pollution and, in the absence of any direction from the DoE, he would exert the power of DAFF on local authorities to refrain from prosecuting farmers for water pollution offences. The level of support the Minister was willing to extend was conveyed in his speech to the Seanad, where he remarked that:

> where section 12s have been issued prior to 27 April, I would be looking to give them conditional approvals irrespective of their date and my Department would be sympathetic to those cases. In view of the overall situation, *I am anxious that local authorities will be sensitive to this matter and I will ask my officials to communicate to them in this regard* (Yates, *Seanad Debates*, 1996, vol. 143, p. 956; emphasis added).

Clearly, while the Minister for the Environment claimed that the collapse of the CFYPS did not come within his remit, and did not therefore require comment, the sentiments of the Minister for Agriculture, Food and Forestry were rather different. Alarmed at the prospect of local authorities pursuing errant farmers for pollution offences after the collapse of the CFYPS, the actions of the Minister for Agriculture, Food and Forestry provide palpable evidence of the importance of political access to the IFA. This is not about 'handshakes' or shady deals brokered in the corridors of Leinster House, but about the success of the agricultural policy community in engineering a consensus on the vital role of agriculture in the Irish polity. The ability to gain access has allowed the IFA to play a significant role in the construction and maintenance of uniform policy, sustain agendas and, where necessary, ensure that objections receive a favourable hearing.

The proliferation of abbreviations in agri-environmental policy (SACs, NHAs, SPAs etc.) could be easily misinterpreted as a signal of significant reform, that a combination of the EPA and EU Directives had achieved the radical overhaul of a regulatory framework for which environmentalists had been striving. However, the sequence of events in agri-environmental policy in the 1990s provides eloquent testimony to the fact that this part of environmental management displays more in the way of continuity than change with the 1980s. It is clear that the agricultural policy community has been successful in maintaining a largely self-regulatory environmental regime and, when challenged, has successfully thwarted attempts at a radical overhaul, sustaining a situation in which voluntary compliance predominates.

This enduring strength is cogently encapsulated in the views of the Minister of the Environment during the passage of the Waste Management Act when he stated that, attempts to subject 'slurry spreading' to an IPC licence:

> would not be appropriate or practical ... we must be realistic about that. The *agricultural community would not be able to bear that type of imposition* (*Dáil Debates*, 1995, Vol 460, p. 85; emphasis added).

Despite waning public support for agricultural subsidy, the agricultural lobby remains well organised, well connected and, more importantly, institutionalised within the policy process. When difficulties arise, or when trouble is brewing, the buffalo can be heard running.

That the farming function has changed is undeniable. The reform of CAP and new Directives from the EU has undoubtedly shifted the focus of concern among farmers. It is in many ways a shift reflected in the lexicon of modern farming where headage payments, SACs, NHAs, SPAs, and Ewe Premium schemes have assumed an everyday part of modern agriculture. And yet, on closer inspection it is a terminology which serves to mask rather than reveal. New schemes may make the headlines, reinforcing the impression of change, but a more detailed scrutiny of policy reveals the success of the agricultural lobby in improving income transfers, largely at the expense of environmental conservation.

BATNEEC, IPC Licensing and the Environmental Protection Agency

Reflecting the political mood to embrace change at the end of the 1980s the introduction of BATNEEC and IPC were presented as the two pillars upon which the credibility of a new regulatory framework would stand or fall. The concerns of the environmental lobby had been heard, problems addressed and government was confident that the combination of BATNEEC and IPC licensing would restore Ireland's green credentials. It would, above all else, effect a new departure in Irish politics, introducing a fully integrated agency with transparency to the fore. It was a shift in policy which acknowledged the need to eschew previous regulatory practices and install a new 'integrated ethos'. The coalition Government was also insistent that the formation of the EPA would provide a basis from which to establish a new environmental accord with business.

IPC licensing was the linchpin of the agency's functions, designed to introduce better standards and ensure compliance. In the process of securing (and retaining) an IPC licence, it was hoped that business would take more seriously the environmental dimension to production practice (Taylor, 1998b). Yet, the combination of IPC and BATNEEC is more than simply another set of procedures; it is, as Duffy observes, more than compliance, it is compliance *plus*, a guiding philosophy which contains a discernible hierarchy; elimination, reduction, recovery, treatment and finally disposal (Duffy, 1995, p. 88). Moreover, it is a philosophy which aspires to an on-going programme of innovation, a process in which environmental management constantly seeks improvement and attempts to anticipate environmental problems (EPA, 1994). Indeed, the EPA BATNEEC guidance notes expressly state that the aim of IPC is to eliminate or minimise the risk of harm to the environment by preventing the emission of potentially polluting substances wherever it is practicable or to minimise such emissions where it is not practicable (EPA, 1994, p. 2; Taylor, 1998b).

Attuned to the difficulties of the 1980s, where environmentalists were abhorred by the proliferation of meetings 'behind closed doors' by agencies such as the IDA, the new regulatory regime places an emphasis upon the need to ensure the dissemination of environmental information. The intention was that important environmental information, from compliance and release monitoring data to detailed justifications of business practices would, for the first time, be placed in the public domain. Public registers would be the forum in which transparency was secured and oral hearings the arena in which disputes or challenges would be enacted (Taylor, 1998b). The decision either to grant or refuse a licence is made with reference to BATNEEC. As its guidelines point out, the technology (which extends to production techniques) would be 'best at preventing pollution, and available in the sense that it is procurable by the operator of the activity concerned'. NEEC would set out the 'balance between environmental benefit and financial cost' (EPA, 1994, p. 2; Taylor, 1998b). Simple, straightforward and above all, intelligible.

To its critics, however, BATNEEC is flawed, a regressive step from the EU's attempt to use the Best Available Technology (BAT) as the basis from which to make judgements on licences. Ultimately, the problems in BATNEEC are located in the inherent tension which resides in the two elements to the equation; BAT and NEEC. This is not immediately clear from a cursory reading of either the legislation or the guidance notes, due largely to the discrepancy which occurs between legislative intent and regulatory practice.

From the outset, it is important to recognise that BATNEEC is not 'cast in stone'. Far from it. Indeed, the agency refuses to be drawn on specific standards or absolute emission levels, opting to treat each licence application on 'its merits'. It defends this position by highlighting the evolutionary nature of the IPC process, wishing to avoid 'fossilising technology options in rigid official documentation' (Meehan, 1994).[22] As such, environmental limit values (ELVs) in the BATNEEC guidance notes are 'not legally binding', more akin to performance criteria linked directly to BATNEEC (Meehan, 1994, p. 79). Although at times BATNEEC can appear disarmingly simple, in practice it can be quite the opposite. In the UK it was noted that there was a conspicuous omission on how environmental risk would be defined, assessed, measured and monitored. On this matter, one of the HMIP's Directors confessed that earlier forms of guidance had 'failed to put flesh on the ideas'. Its response has been to establish the overall environmental impact of a process quantified in the Best Practicable Environmental Option (BPEO) Index. The process option (either the practice or technology) adopted would be given an index rating. If the operator wishes to use the BEO for the company then no further assessment will be required. If not, then a full justification on the grounds of NEEC is required. SACs and NHAs are covered by ministerial regulations which provide for development in instances of 'overriding health and safety' or for 'imperative reasons of overriding public interest'. As Hickie notes, some large scale ventures, such as sewage treatment plants or landfill sites, could prove controversial (Hickie, 1997, p. 28). On closer inspection, however, there are tensions which reside within BATNEEC which have important consequences for the regulatory practices of the EPA.

It is, for example, commonplace for the agency to make a distinction between existing (as opposed to new) facilities, one which may allow the agency to stipulate more 'lenient' conditions on the best available technology for existing or older production processes because of the costs incurred in any attempt to overhaul the production regime (Taylor, 1998b). This has at least two important repercussions. First, the creation of established activity status means that most existing First Schedule activities will be given a number of years to come up to the standards of new facilities. This confirms the fear among environmentalists that the standards required under BATNEEC are not as stringent as those required under BAT. Indeed, on this matter, one of the EPA's Directors pointed out that there was little difference between and BAT and BATNEEC for new activities. However, he also noted that while existing activities would have to meet BAT under the directive, they would do so only in 11 years time (Maclean, 1994, p. 82). Second, the Minister possesses significant discretions regarding the time at which a class of operation should become licensable. This was clearly an important factor in the ability of the agricultural lobby to delay the point at which intensive pig and poultry farms came under the jurisdiction of the agency.

It is essentially the NEEC element to the equation which attracts controversy. It would hardly be stretching the bounds of the imagination, for example, to expect a strong lobby group such as IBEC or the IFA to extract concessions on licence conditions on the grounds that they 'entail excessive cost'. On this issue, the agency has acceded that it is willing to go 'as far as possibly to accept what an applicant company believes is excessive cost' (Taylor, 1998b). Remember, the agency's function is not to prevent installations operating, it is to create a compromise in which practices are deemed acceptable. This is the crux of the problem. What is an acceptable practice is a politically charged concept, always in a state of flux and open to considerable variation. It is necessarily dependent upon the level of economic growth, the global economic climate, the vulnerability of individual sectors of the economy and the capacity of the environment to absorb pollution.[23] What is more, the NEEC element to the equation must also consider the power of multi-national companies to reject the EPA's stance, which may lead to production being shifted to more favourable locations (pollution havens). To date, the EPA has avoided the inclusion of global pollution processes (greenhouse effect) within calculations of BATNEEC, preferring to concentrate on the local dimension, presumably on the assumption that if local problems are resolved the big picture will look after itself.

In a more contentious tone, critics of the EPA suggest that the BATNEEC equation is presented in a quasi-scientific manner, as if it is simply a case of adding up the advantages and disadvantages of a particular technology before arriving at an acceptable practice. Objective, precise, neutral and performed by those 'who know', it is an anodyne process, isolated from the messy world of politics and, above all else, downplays the fuzziness of the NEEC element to the equation, finding solace in procedures which favour taking 'each case on its merits'. And yet, what is an acceptable practice today may not be acceptable practice tomorrow. For BSE today ('an event which couldn't happen', and if you were a scientist who thought it could, then you were a maverick) read genetically modified food, or spreading blood and animal by-products on land. The term acceptable practice is inevitably influenced by the state of scientific knowledge, the weight of public opinion or, perhaps more ominously, the changing balance between environmental conservation and economic growth. It is, therefore, a term which is irrevocably political, and not surprisingly, covers matters not dealt with in the guidance notes.

In general terms, these difficulties arise because of the discrepancy which emerges between legislative intent and regulatory practice, repeated in the area of licence enforcement where the EPA has declared that its role is not to document all breaches of licence conditions; rather, minor variances are *noted* and the company is then informed. It is only in cases of larger variances that an investigation is pursued through inspection. Only in circumstances of *significant* variances are prosecutions sought (Maclean, 1994, p. 82 cited in Taylor, 1998b). Such a style of enforcement leaves the EPA vulnerable to the charge that its relations with industry are 'too cosy' for it to operate its vigil competently.

This was a problem which had also been encountered in the UK at the end of the 1980s where the public credibility of Her Majesty's Inspectorate of Pollution (HMIP) was at a low ebb. Criticised for a low level of enforcement action, site inspections were at the nub of the issue. Indeed, one company suggested that they amounted to little more than an inspector popping in 'for a cup of tea'. The introduction of IPC, with the intention of instigating an arms length relationship was not without problems. In particular, industry felt that HMIP's:

> withdrawal from its traditional role of advising industry about projects at an early stage meant that an awful lot of time could be spent preparing an application ... only to be told to go away and do it again (ENDS, 1994, p. 16).

On a more positive note, the EPA has shown a welcome zeal for disseminating environmental information on monitoring, and has initiated a succession of innovative methods for the collation of information. These include annual environmental reports (AER), pollution emission registers (PER) and environmental audits. Although still at an experimental stage these techniques should have the potential to enhance both the level of environmental information available to the public and provide an environmental snapshot of particular licensees. The PERs for example, trace the breakdown of chemicals through the production process, providing invaluable information on how many raw materials are recycled and how much ends up in the final product. It may form the basis for a change in the way licensees are able to identify potential areas for reducing waste chemicals. Moreover, it should allow comparisons to be drawn both across geographical boundaries and between industry (Taylor, 1998b).

On a more disappointing note, only 23 PERs were submitted, with the majority coming from the pharmachemical industries in Cork and Dublin. At this juncture, judgement should be reserved, since many facilities do not currently have the necessary in-house expertise to complete such formats. The EPA remains confident however that, with time, improvements in submissions can be achieved (EPA, 1997, p. 15). These methods of data collation are complemented by inspections and audits conducted by the EPA. In 1997, the EPA conducted 1439 visits (889 for sample emissions and 550 for auditing and inspection). However, of 387 IPC licensees only 60 were audited, and among these the EPA's research reveals a disturbing pattern of non-compliance (48). The most significant lapses were for document control and procedural irregularity. Here, the EPA's report noted that at 'these facilities responsibility for issuing and revising procedures was not set out, nor were procedures, where issued, available to the relevant operatives. Moreover, 17% of sites recorded 'poor document control' and in some facilities 'documentation could not be retrieved upon request'. In many instances, files were 'present in multiple locations, out of date, or staff were unaware of their existence' (EPA, 1997, p. 19). What is more the EPA's findings confirm that many reports required in the IPC licence submitted to the agency did not contain all the results specified, or that monitoring was conducted in the right manner for the frequency specified by the agency (Taylor, 1998b).

However, by far the most disturbing feature of the EPA's report is the picture it portrays of consistent non-compliance. In 1997 the agency received 4044 reports, of which 20% were non-compliant, while 9% had still not been submitted by the end of 1997. A further 11% were outstanding and under assessment by inspectors. The report also noted that non-compliance had increased significantly over 1996.

The problem the agency faces is that by adopting a stance which avoids punitive action, in the hope of securing greater awareness and thereby developing environmental standards, the legitimacy of the regulatory framework may be brought into question. For many environmentalists, this is aggravated by the fact the agency is willing to assist operators through all phases of the application procedure.

This is clearly an important element to IPC, one in which regulatory authorities seek to achieve a balance between imposing licence conditions (often new and many of which business will be unfamiliar with) and the need to co-opt business into a system which is non bureaucratic in order to encourage compliance. It is by no means an easy task. In the UK, where IPC licensing had been established as part of the 1990 Environmental Protection Act, Her Majesty's Inspectorate of pollution (HMIP) was keen to impose an 'arm's length relationship' in order to establish a regulatory distance between the regulator and those being regulated, signalling an effort to end an allegedly 'cosy' relationship with industry. HMIP declared that it was reluctant to engage with industry in pre-application procedures or offer advice, rejecting the possibility of becoming an 'environmental consultant' (Taylor, 1998b). In practice, however, the first round of IPC licence applications proved tortuous, leading to excessive delays. HMIP's initial reaction was to hold 'off the record' meetings, a move which undermined the transparency of the system. The policy of an 'arms length relationship' was formally dropped in 1993 (ENDS, 1994, p. ix). It is worth noting that despite the influence of British environmental policy in general the EPA's stance on this issue has been markedly different. The EPA has expressed the view that in order to avoid possible delays 'requests for discussion' would be 'favourably considered' (EPA, 1994, p. 7).

There are clearly areas in which the introduction of the EPA has made important (and beneficial) improvements in environmental regulation in Ireland. The introduction of new techniques for the collation of data and monitoring bode well for the future. Moreover, the agency has displayed a welcome commitment to provide environmental information in a wide variety of areas. However, problems remain. In particular the high rates of non-compliance and a reluctance on the part of the agency to pursue punitive action are a source of concern. In part, this can be attributed to the fact that the agency is forced to use District Courts, an avenue which may deliver judgements quickly but one less inclined to impose large fines on serious (and serial) polluters. Innovative and ground breaking though some of the measures are, it does not remove the 'feeling' that perhaps its relationship is rather too accommodating and that this certainly contrasts with its view towards third party objectors. An issue we will return to in the next chapter.

Waste Management and the Environmental Protection Agency

The problem of waste management has emerged as one of the most politically contentious areas of environmental politics in Ireland. Disputes such as those at Kill dump in Kildare and the controversial closure of Carrowbrowne in Galway, have ensured that landfills, Not in My Back Yardism (NIMBYs) and debates about recycling remain an increasingly common feature of the Irish political landscape. With little attention given to location and more often than not poorly managed, landfills have been a consistent source of local protest. As if to complicate matters further, it now seems that a 'generation' of landfills, extended beyond their anticipated operational lifespan, are on the verge of closure, a situation which could rapidly precipitate a crisis in waste disposal.

This has emerged as a serious concern among residents in Silvermines, Co Tipperary where a subsidiary of Waste Management Inc (a huge American multi-national company) has proposed to develop a disused barytes mine containing 1.6 million cubic metres of toxic waste into a landfill. The *Irish Times* reported that a number of local authorities were 'waiting to see' whether WMI would get a licence before deciding what to do with their waste (*IT*, 25 Jan 1999).

It is a debate both politically charged and complex, one which certainly does not revolve simply around the estate agent's old adage that what is important is location, location and location. Here, few would disagree that a policy vacuum has emerged, largely because of a failure on the part of central government to define the parameters of policy. Local authorities are faced with the prospect of making decisions not only on the location of landfills, or on whether to transport waste to other areas (super dumps), but on how to finance waste disposal. Contentious though this area of environmental regulation is, it is depressing to note that the EPA's role remains far more circumscribed than many had anticipated. Indeed, it is an area of policy which reveals, once again, that 'grey areas' persist, and that integration has been a term loosely defined and often inappropriately adopted in describing the EPA's role.

During the 1970s and 1980s waste management was largely controlled through public health statutes and Ministerial regulations 'implementing relatively undemanding European legislation' (Meehan, 1996, p. 59; Scannell, 1995). However, from the mid-1980s onwards it attracted the attention of EU environmental policymakers. As a consequence, the government responded with the Waste Management Act, 1996 which attempted to draw together many of the objectives of EU policy. As such, the Waste Management Act accords the EPA a central role in regulating waste, places a duty upon it to produce a national hazardous waste plan and grants the agency powers to issue and monitor licences for the disposal and recovery of waste (Meehan, 1996, p. 59).

The Act contains many policy developments synonymous with recent EU Directives and incorporated into the EPA Act, 1992; a preference for management plans, licensing, compliance monitoring and significant discretions on how to achieve the goals set. As such, the Act affords regulatory bodies a variety of options, ranging from charges and taxes to the dissemination of information and operation of voluntary agreements (Meehan, 1996, p. 59). To date, the most prominent development in this regard has been the agreement struck with IBEC over waste recycling whereby the Minister set up a voluntary agreement. It is one of several avenues available to the Minister who may stimulate recycling by providing financial assistance, establish voluntary programmes, mandatory regulations or by a combination of all three.

In the case of packaging waste, successive administrations have been prepared to give industry the opportunity to organise voluntary approaches to the problem suggesting that, in the absence of satisfactory progress, mandatory schemes would be imposed. IBEC responded with its recycling and packaging initiative that set modest, but comfortably achievable targets (Meehan, 1996, p. 60). However, by far the most significant area of change in the area of waste management in Ireland has been the introduction of waste licences regulated by the EPA. Here, section 98 of the Act has redefined the relationship between the EPA and the local authority. The Act stipulates those operations which require an IPC licence for the incineration of hazardous and hospital waste, waste with a capacity of more than one tonne per hour and the use of heat for the manufacture of fuel from waste (Brassil, 1996a). It would be churlish to suggest that the Waste Management Act is not a significant improvement on previous regulations. However, this should not detract our attention from problems which persist, particularly with regard to landfill charges and the designation of sites.

One of the more controversial aspects of waste management in Ireland has been the dependence upon landfill, a method of disposal which accounts for 70% of commercial waste and 98% of domestic waste. This preponderance on landfill is aggravated further by a failure to establish a national waste policy. Successive administrations have chosen to leave decisions to local authorities, defending this position on the grounds of the EU's preference for the principle that the polluter should pay. This means that the cost of collection and disposal of waste, either from industrial, commercial or domestic sources, should be met as close to the source as possible. How local authorities are supposed to devise the financing of these schemes remains unresolved. Landfill charges have been mooted, a policy option favoured in some quarters (see Barrett and Lawlor, 1996). However, politicians are innately conservative creatures and are haunted by the spectre of being the harbinger of an 'Irish poll tax'.

In the UK, the poll tax replaced local authority rates. It was a charge made to each individual (as opposed to a household). The Conservative Government argued that since each individual consumes services equally, everyone should pay the charge. It seemed incomprehensible to Mrs Thatcher that this was not self evident. Political debate, however, became hamstrung by the issue of regressive forms of taxation. Examples of individuals living in mansions paying less than those living in council houses were common currency. The irony in this was that the poll tax was originally conceived as a mechanism to usurp the last bastion of socialism in Britain: local government (or at least local authorities under Labour control). The intention was that the poll tax would (re)create voters as consumers (both of local politics and local services) thereby ensuring a link between individual consumption and voting preference. Mrs

Thatcher remained convinced that the electorate would finally recognise that Labour controlled authorities were profligate, extravagant and inefficient. The problem was that everyone seemed to have their own little horror story, or ludicrous anecdote, about houses in the same street being charged different levels. I mention this because if either water charges or refuse collection charges are introduced in Ireland, controversy will occur over any variations which emerge around the country.

In theory, such a charge should stimulate a diversion of waste up the hierarchy, encouraging composting, recycling, waste minimisation or incineration with an element of energy recovery. To date, there has been little in the way of a serious commitment to recycling, composting or waste minimisation around the country. The only example of a domestic kerbside collection system is a pilot project in Dublin. However, as Barrett and Lawlor point out, price volatility in this area presents a significant problem to its economic feasibility. The average price of recyclables in Sept, 1993 was £17 per tonne, a figure which increased to £60 per tonne in June, 1995 but then returned to £30 per tonne (Barrett and Lawlor, 1995). Barrett and Lawlor estimate that aluminium recycling is profitable in its own right, whereas the recycling of bottles and textiles depends on large volumes and sustained prices (Barrett and Lawlor, 1995, p. 76 and 1997).

The designation of landfill sites is a further area of concern and raises serious questions about the division of responsibilities between the EPA and the planning authorities. While it is plausible (if not entirely convincing) to suggest that the 'grey areas' which materialised in the EPA Act were an 'oversight', no such defence can be offered with regard to the Waste Management Act. It was an issue raised extensively at both the committee stage of the bill and the final stages in the Dáil. Trevor Sargent of the Green Party, for example, moved an amendment which stated that:

> a planning authority will consider a planning application for a development that requires a waste licence *only after that licence has been granted* (*Dáil Debates*, 1996, vol 460, p. 88; emphasis added).

The intention of the amendment was to explicitly avoid the problems raised in the Masonite case. The Minister, while 'mindful of the arguments advanced', conversant with the possibility 'for confusion' and aware of 'anomalies', chose to reject the amendment. Presented with an opportunity to address what was widely acknowledged as a serious deficiency in the EPA Act, the government stalled, sheepishly looked around and found cover in the bemusing phrase that it wanted to establish an:

> expert body on the environment for Ireland and did not want to alter sub section (3) of the Act which delineates the respective roles of the planning authorities and the EPA (*Dáil Debates*, 1996, vol 460, p. 88).

Why? The Government's abject response was that by reversing the EPA's position in the 'queue to assess' it would unnecessarily crowd the planning arena.

For those more sceptical of the Government's green credentials, there are alternative arguments, found not in the role of the EPA, but in planning procedure. After all, the government seemed content with the success of changes to planning procedure introduced in 1992 which had increased the speed with which An Bord Pleanála makes a determination. Put simply, if it ain't broke, don't fix it. Or, could it be that a two stage process, where neither regulatory body is there to 'prevent development', is more conducive to the success of potential projects? It is an argument which is persuasive only if we recognise that this is an era in which encouraging the entrepreneurial spirit is to the fore and that, any move which could undermine that most fragile and elusive features of the Emerald Tiger, business confidence. It is certainly the case, as we shall explore later, that

Government perceives the agency to be a 'body of experts' where research, advice and environmental consultancy are its most appropriate roles.

The tension which exists in the sub-division of responsibilities between the EPA and the planning authorities reappear in the area of mineral extraction which has been designated a schedule 1 activity and is subject, therefore, to EPA licensing. It should be noted however that, unlike most other list 1 activities, mineral extraction and storage is 'unique', in that the conditions of the licence may never lapse; the mine may have closed, but the storage of waste and post-closure care present innumerable difficulties for future environmental management (J Derham, 1995).

While the scarring of the landscape is undoubtedly the most visible environmental intrusion, one which usually attracts the more animated forms of political protest, policy also needs to address some of the more complicated issues which surround mining and which are aired only rarely in the public domain. The recent pollution incident at Silvermines, Co Tipperary is a case in point, where waste storage, acid generation and the release of toxic substances into groundwater supplies emerged as important features of a lapse in post-closure care. The problems presented to the EPA in this area are borne largely out of the sub-division of responsibilities with the planning authorities and the complexities surrounding the transfer of mine ownership and post-closure care. To most people, including this author, it would have seemed appropriate to include the developmental work of the mine within the environmental evaluation for licensing undertaken by the EPA. A cursory reading of the legislation would suggest that the EPA's powers in this field are extensive. However, on closer examination the failure to assign a regulatory role to the EPA with regard to developmental work at a mine once again raises the issue of 'grey areas' of responsibility between the EPA and the planning authorities. The EPA emerges as the competent authority only when the company is up and running or, in this case, when a product is being extracted. However, an important distinction needs to be drawn between cases such as Masonite and that of mineral extraction, since mining carries the potential to raise pollution matters 'further down the road'. Excavated material has to be stored, and disposal therefore needs to be controlled (Derham, 1995; Doyle, 1996).

It is not the only area in which the EPA has experienced difficulty as the legislation has struggled to come to terms with the complexity of changing ownership and post-closure care, issues which came to the fore in the investigation surrounding problems at the Tailings Management Facility at Silvermines in Co Tipperary where mine tailings waste was produced. This is a waste from a metal ore processing mill following the extraction of metal concentrate from finely crushed rock. The risk posed from such waste depends to a large extent on the management methods and standards applied. The primary ores extracted at Silvermines were lead, zinc and silver (which contained pyrite and chalcopyrite). The latter two are high sulphide minerals and when incorporated into tailings waste have the potential to become acidic and produce acid rock drainage. Such waste storage requires perpetual maintenance (EPA, 1999). In instances where underground storage is preferred, and where groundwater has been lowered, consideration should be given to the potential for a 'shock load' of contamination arising from back-filled waste when the mine is re-flooded (EPA, 1999). Once flooded, the on-going risk of acid generation is reduced and thus enhanced mobility of priority contaminants is inhibited. Attention should be paid, therefore, to removing any potentially polluting process chemicals from the tailings waste prior to backfilling, otherwise there is a high risk of migration (Derham, 1995).

Tailings waste disposal is usually stored in varying proportions between underground disposal and surface impoundment. As Derham points out, the principal driving force behind the agency's approach to mining waste is the

EU Groundwater Directive (80/68/EEC) which prohibits certain List 1 substances and requires the limitation of others (List 2) from entering groundwater. The agency is required to ensure that all technical precautions are taken to prevent the discharge of List 1 substances. The regulations (Local Government and Water Pollution) 1992 contain a zero quality standard in respect of List 1 substances. However, this presents the 'EPA with a legal difficulty in that there is no laboratory that can determine a zero value in any analysis' (Derham, 1995, p. 130). Where surface storage is chosen, water saturation of the waste has to be maintained and presents a residual and long-term pollution potential long after the mine has stopped producing. As Derham notes, 'this means that all surface disposal taking potentially polluting mine wastes may be required to have a composite lined base (flexible membrane liner on a mineral layer)'. Some geological and hydrogeological conditions may permit departure from this standard. The agency has been looking into design requirements for the disposal of non-hazardous industrial waste in facilities across Europe and is considering adopting the EU Landfill Directive 9570/95 ENV 185, 19,9,95 (Derham, 1995, p. 130).

In the case of the TMF at Silvermines, which the EPA reported as being 'a perpetual risk to health and the environment', the pollution stemmed from a tailings pond which had been rehabilitated in the mid-1980s. In one instance a release of toxic dust clouds forced some local residents to evacuate their homes. The mine was formerly owned by Mogul Ireland and closed in 1982. The company itself was taken over by Ennex International in 1984, and the Tailings Management Facility was later sold to a farmer with no mining or waste management experience, a move which the EPA believes contravenes the Waste Management Act, 1996 Section 32(2) (EPA, 1999).

The matter of waste storage in cases such as this presents considerable difficulties to the EPA who consider that, 'storage of mineral waste is only subject to a IPC licence where *related* to IPC'd extractive and processing operations' (Derham, 1995, p. 129). 'Related' in this context could mean:

> either part of the same operation (carried out by the same company operating the processing plant) or it could mean waste storage by a separate company on contract where the waste originates from an IPC'd extractive operation (Derham, 1995, p. 130).

Put simply, it becomes an issue of establishing who owns the site, who is responsible and, ultimately, who pays.

In certain areas of planning there is legal provision which creates a financial guarantee or bond, and is operated by the Department of Trade, Energy and Communications and the planning authorities. However, planning bonds of this type tend to be finite and are usually exhausted or returned to the operator on the completion of the project (Derham, 1995). It is one thing to set aside a fund to deal with a defined closure plan, where costings can be confidently approximated, quite another where remediation is required in the event of unforeseen circumstances. It is a matter upon which the EPA has made important strides, rectifying previous policy failures and should be commended.

At the European level, the issue of environmental liability emerged as a logical extension to the 'phasing in' of the polluter pays principle. The intention was to put in place a liability regime which would encourage the implementation of environmental rules. As the EU Commissioner on environmental liability stated:

> it is essential we have a system in place which will force the people who cause pollution to clean it up. Even more important, we must ensure that everyone engaged in potentially dangerous activities knows about the risks of being liable in this way, so that they have a strong incentive to prevent pollution happening in the first place (Bjerregaard, cited in Derham, 1999, p. 2).

One of the obstacles to developing a policy on environmental liability is that very often public liability policies have a clause which excludes coverage in areas thought likely to incur large costs. As Derham notes, generally what is covered is the 'big bang event', rather than long-term environmental damage. However, under regulations from 1996 the EPA is empowered, where it thinks necessary, to require IPC licensees to furnish information on financial security to meet commitments or liabilities which may arise after the operation has ceased production (Derham, 1999). It is a welcome development and should reduce the possibility of incidences such as that at Silvermines.

Once again, the EPA's role in many areas of waste management represents a distinct improvement, most notably in the introduction of monitoring and IPC licences for landfill sites. However, the persistence of grey areas of responsibility present significant hurdles to be overcome. It is a reflection not so much of any shortcomings in the EPA, as the unwillingness on the part of government to create a legislative framework which could have resolved these difficulties. It is a political issue, an omission which has its roots in the ability of the planning lobby to persuade government that a 'streamlined' planning appeals system is essential to the health of the Emerald Tiger.

Planning and Environmental Regulation

Although strictly speaking planning is not an area in which the EPA has a significant role it does have an important bearing upon the context in which the EPA operates. It is certainly an arena which has witnessed some of the most sophisticated and politicised forms of environmental protest in Ireland. Lancefort (a non-commercial company that aims to raise crucial issues on the role of the planning process), for example, has sought to challenge a number of prominent planning applications through litigation, a move which sparked the comment that developers were 'apoplectic at never knowing when they were going to be ambushed' (see S. Dillon, *Irish Times*, 27 March, 1998). Such an approach can be contrasted with the tactics of eco-warriors in the Glens of Wicklow, where this form of opposition to the extension of the road network is a new phenomenon on the Irish political landscape.

During the late 1980s the construction industry, which has experienced a sustained speculative boom in the last eight years, persistently sought changes to a planning system felt to be excessively protracted and cumbersome. It was an argument which found sympathy in Government circles, where the prevailing view was that planning should no longer be regarded as a purely local or regional matter. As the Minister noted, delays in the:

> planning system, or any perception that our system is ponderous or dilatory, ... put us at a serious competitive disadvantage in seeking to attract internationally mobile projects (Smith, *Dáil Debates*, 1992, vol. 416, p. 73).

It should come as no surprise then that amid a speculative property boom such arguments are now in the ascendant. If demand is to be satisfied then supply needs to be increased. The construction lobby has convinced government that this can be achieved only with a combination of rezoning and quicker planning decisions. These are emotive issues linked principally, although not exclusively, to Ireland's housing policy, which has eschewed the forms of tenure associated with France, Germany and Scandinavia in favour of home ownership, a policy it shares with the UK tax reliefs on rental income, inner city building, seaside resorts and mortgage interest relief (in other words discriminatory policies which favour home ownership and speculative property development) have been pursued at the expense of policies which favour integrated policies which allow greater household choice in terms of tenure (see Balchin, 1996). Amid the political furore generated by developers during this period, often bordering on the hysterical, it is hardly surprising that the decisions of An Bord Pleanála should come under intense scrutiny. As far as the planning lobby was concerned planning procedures were time consuming, protracted and, ultimately, costly. Change was imperative if the construction industry was to service adequately the Emerald Tiger. In defence of An Bord Pleanála, it should be recognised that the board had been denied any new funding or staff for a considerable time and, although appeals to the Bord had increased by 35% in the period 1988–1989, general staff had been cut by 20% and the crucial grade of inspectors by 50% (Howlin, *Dáil Debates*, 1992, vol. 416, p73).

The introduction of the Local Government (Planning and Development) Act, 1992 was the government's response, and aimed to 'streamline' the appeals stage of the development process by focusing on the need to reduce the time period for a Bord Pleanála decision on an appeal. The impact of these changes has been considerable, and perhaps most visible in the time it now takes for the Board to make a determination on appeal. In 1992, 34% of appeals were dealt with in the four month period and yet, by 1994, the board handled 2,330 appeals of which 98.5 % were completed in the four month time limit.

In response to its critics the Bord argued that, while it was fully aware of the pressure to process appeals more efficiently, many appeals are:

> probably cases that have raised difficult issues at local level arising from the implementation of the provisions of the Development plan, the protection of amenities and the need to reconcile proposals with the proper planning and development of the area. A number raise contentious third party issues. These cases by their nature require more time for processing and careful consideration, a factor which is sometimes overlooked in criticisms of the system (Bord Pleanála, 1995, p. 5).

The Act introduced a 'one shot' appeal system whereby all grounds of appeal had to be submitted within a time limit of one month (Galligan, 1997, p. 237).

Under previous regulations, those making appeals could submit initial grounds and then elaborate at a later stage. In such circumstances, it was not uncommon for an 'on going exchange of documentation' between the Board and those making appeals to take place (Galligan, 1997, p. 237). The Act also contained two significant changes to the regulatory process which had an important bearing upon the democratic nature of environmental regulation. First, it extended the time period in which third parties could lodge appeals. Second, and perhaps more importantly, it stipulated that objections would have to be made *in full* during this time period.

At a cursory glance the extension in the time period allowed for appeal would appear to have improved the democratic nature of the planning process. However, it is an illusion soon exposed when it is recognised that no extensions to the original objection are permissible in the light of new conditions after the time period has elapsed. In other words, in complex cases, where (additional) information is either required or not readily available or, where experts need to be consulted, the quality of the objections (and presumably their potential success) are severely curtailed. The objection must be made in full, no further elaborations are permissible and the appellant must state whether s/he intends to make a request for an oral hearing (Galligan, 1997, p. 240). With such tight technical regulations it is hardly surprising that the 'enterprising' should seek to gain advantage. As Galligan has observed, it is not uncommon:

> for developers to lodge an appeal at the eleventh hour so that a third party who is slow to react may find himself excluded due to the operation of the statutory time limit (Galligan, 1997, p. 242).

In many cases it may be prudent for third party objectors to make 'observations'. Although not entitled to request an oral hearing, the third party would still be able to participate fully in the planning process and any subsequent oral hearing requested by another party (see Galligan, 1997, p. 242).

While the 1992 Local Government and Planning Act does not legally remove any right of appeal from third parties, there is no doubt in the minds of most commentators that the ease with which this right can be exercised has been diminished. The argument that the 1992 act is an erosion of public participation in the system is further compounded by the massive increase in the appeal fee for a third party to £120. It is also disturbing to note, as Galligan does, that there is no regulation which allows a third party objector to recover costs which have been incurred in the preparation of a planning appeal (Galligan, 1996b, p. 142). The size of such costs can often be considerable and must surely represent a significant reason for undermining 'potential' objectors.

In situations such as this there is the very real prospect that the new appeal stipulations will clearly leave only well resourced developers or well resourced communities as players in the planning process. Several developers have recently expressed concern that a number of vexatious or frivolous appeals still impede the development process. Pat Nolan (Hamilton, Osborne Kings) has argued that the fee should be increased in relation to the size of the proposed development. As a result 'objections to large developments would come from housing associations rather than from any person on every street corner' (*Irish Times*, 24 June 1998). In its eagerness to dismiss frivolous or vexatious appeals, the legislation may undermine the role of protest from individual citizens, resident associations or communities altogether. It certainly represents a considerable shift in the spirit of planning law which has, since the enactment of the 1963 Act, been in favour of encouraging participation.

The Minister insisted that the existing rights of third parties to appeal would in no way be diminished and that the aim was to 'create a more orderly and effective procedural framework' (Smith, *Dáil Debates*, 1992, vol 416, p. 72). In his statements to the House, however, it was clear that it was not the rights of third party objectors which were foremost in his mind. While he recognised the need for a 'rigorous examination' of planning proposals it was, in his opinion, incumbent upon the Minister to be aware of a developer's entitlement to expect a final and conclusive decision within a specified period of time. Undue delay would cause 'understandable frustration' for developers, with a 'consequent loss of investment and employment opportunities' (Smith, *Dáil Debates*, 1992, vol 416, p. 73). Certainly, if the enthusiastic welcome the legislation received from the Irish Construction Industry Federation is anything to go by, then the balance has tilted firmly toward the developer.

A further area within planning which has received attention has been the introduction of Environmental Impact Assessment (EIA), which for many environmentalists was anticipated to provide more rigorous control of agriculture. It was a sector which had been subject only to a limited form of environmental control under EC Regulations. Following the introduction of EIA there is now a statutory obligation on the part of the planning authorities to review the content of an EIS and the legislation allows for third party objections and comment on the same (J. Fry, 1996).

In more recent years, however, attention has focused on the quality of the EIS's submitted where the conclusions have tended to be rather more suspect than the views expounded earlier by either the Environmental Research Unit (ERU) or Scannell, who, at the time, were cautiously optimistic. The observations of both the ERU and Scannell were made very early in the experience of EIA. Even now there is, as Fry notes, a lack of an inbuilt audit of cases where EIS's have been submitted and improvements in information are necessary if standards are to be improved (see J. Fry, 1996). Fry, for example, concludes that the standard of EIS's submitted has tended to be unsatisfactory, particularly for those in intensive animal housing (J. Fry, 1996, p. 153). In his opinion, this is confirmed in the propensity for local authorities (particularly smaller authorities) to make requests for further information, an avenue that is generally utilised to offset inadequacies in the EIS. And yet, despite evident inadequacies in the standard of EIS's submitted the research shows that, of the cases studied, 84% of applications were granted. In addition, of the 50% appealed to An Bord Pleanála, 60% were successful, prompting Fry to remark that the 'EIA procedures do not militate against a positive decision for the developer prepared to go through the process' (J Fry, 1996, p. 154). The conclusion drawn by Fry, that intensive animal housing had tended to skew the findings because of the 'unusually poor quality' of submissions in this area has been confirmed in the work of Byrne (1994).

Since the adoption of EIA Regulations, large-scale pig and poultry operations are now required to submit an Environmental Impact Statement (EIS) with the planning application. Byrne's study, however, has shown that the standard of the EIS's submitted has, in many cases, been highly unsatisfactory and that a number of flaws can be identified: no scoping had been carried out with the authorities as to what was required in an EIS; none of them mentioned the construction phase, and only one attempted to describe the site; many of the applications related to units currently in operation but did not mention the impact of existing operations; very few of the EISs referred to any data on surface water quality in the area, and those which did, relied on data at least two years out of date; very few mentioned the impact on ground water, and none presented any data on the subject. Of the 32 applications studied only 'seven could be said to have been reasonably unbiased, with most lobbying for the development to go ahead'. And yet, remarkably, of these 32 applications, 26 were successful (D

Byrne, 1994, p. 124).[24] In Byrne's opinion, the EIS's submitted were of 'such poor quality that it is difficult to accept that the planning authorities could decide on the basis of the EIS alone' (D Byrne, 1994, p. 126).

One of the principal environmental pollution concerns in such forms of intensive farming is the risk of effluent or slurry run-off into groundwater. And yet, while some authorities set 'broad conditions', there was a wide variation on the size of the 'exclusion zone' in the event of run-off. In the case of some surface water streams, such exclusion zones could be as small as ten metres or as large as 100m. The lack of consistency on this matter is a source of considerable concern and has been mirrored in the conditions attached to planning permission, where considerable variation exists in the number and severity of conditions imposed (Byrne, 1994).

Byrne's research also reveals serious deficiencies on the part of the planning authorities. Of the three planning authorities visited, none had kept records of environmental monitoring. One authority, which did not make a practise of inspecting units after construction, was unaware of whether or not some of the units for which permission was granted had ever even been built. The other two said they did inspect, but did not assess whether work undertaken had complied with planning conditions. Given the poor environmental record of this industry it is astounding that one authority could offer the defence that it was in the interest of the developer to build the unit properly. The other two authorities, who claimed that monitoring was required if it was part of the planning conditions, failed to provide documentary evidence (Byrne, 1994, p. 128).

The issue of environmental impact assessment also emerged in the controversial case brought by Lancefort Ltd. Lancefort, a limited company set up by a group of conservationists to pursue planning cases, sought to quash the decision taken by An Bord Pleanála to allow a development proposed by Treasury Holdings. The case raised a number of interesting themes, as well as a novel form of environmental protest. Developers were dismayed that such litigation was costing £60,000 per week and argued that Lancefort, a limited company, had been constructed simply to shield the true applicants against an award of costs (Simmons, 1997, p. 97). The developers protested that Lancefort did not have sufficient interest in the proposal, was insensible to its environment, and should, therefore, not be granted standing to challenge a planning decision (Simmons, 1997; see also Carolan, *Irish Times*, 13 March 1998).

Although the courts did not find in favour of Lancefort, the case reinforced earlier judgements on the *locus standi* of third parties to object to planning developments (Simmons, 1997). Initially, Morris, J appeared to confirm the view taken in *Malahide Community Council Ltd v. Fingal County Council* where:

> good, bad or indifferent planning decisions cannot affect this artificial body in any way, except by increasing or diminishing the asset value of its own lands or buildings favourably or unfavourably affected by such decisions (cited in Simmons, 1997, p. 97).

However, in the case of Lancefort, Morris, J held that the rule of personal standing may be waived if in the particular circumstances 'the court finds that there are weighty countervailing considerations justifying the departure from the rule' (Simmons, 1997, p. 97). Put simply, the entitlement to participate in the planning procedure (and raise objections) derives from public interest or community spirit and not just property rights (see Simmons, 1997 for more detail). For example, in R. v. Hammersmith and Fulham L.B.C ex parte People Before Profit Ltd. (1981) it was argued that:

> a person is entitled in my judgement to object to a planning matter who has a legitimate *bona fide* reason. He does not have to be a ratepayer. He does not have to be a resident. But he must not be an officious bystander or an officious busybody. He must have what any reasonable person would say was a legitimate interest in being heard in objection (cited in Simmons, 1997, p. 99).

The real question, as Wade succinctly observes, rests on 'whether the applicant can show some substantial default or abuse, and not whether his personal right or interests are involved' (Wade, 1994, cited in Simmons, 1997, p. 100).

Some Concluding Remarks

That significant (and sometimes beneficial) change in environmental regulation has been achieved is not in question. The introduction of the EPA Act, 1992, the local Government and Planning Act, 1992 and the Waste Management Act, 1996 have all effected a change in the respective regulatory responsibilities of different agencies in the 1990s. Nor has it been the intention of this chapter to deny that improvements have not been sustained and environmental benefits maintained. On the contrary, with regard to waste management the EPA's role has enabled the adoption of more stringent guidelines, improved monitoring and has implemented more robust procedures for post-closure care and maintenance. Moreover, and in stark contrast to its counterpart in the UK, the EPA has been willing to enter invaluable environmental information into the pubic domain, improving the transparency of the regulatory framework.

However, problems remain. For sure, it would be easy to dismiss the objections of concerned citizens or environmentalists as the views of 'green fundamentalists', fanatics or loose canons. The argument runs that they do not operate in the real world, that it is impossible to police the behaviour of thousand of licensees or, they desire statutory, stringent, intrusive and punitive control simply to 'beat' the EPA and multi-nationals. The problems within such arguments is they espouse little more than a demand for a return to the pragmatic, incremental and loose regulatory style which had failed during the 1970s and 1980s. While important strides forward have been made, difficulties persist, most notably, in the role of BATNEEC, integration, planning regulations and the sub-division of responsibilities between the EPA and the planning authorities. Thus far, assessments of environmental policy have tended to accept the assumption that significant change has been achieved and to ignore the fact that, that in many ways, there is more in the way of continuity than change with the regulatory framework of the 1970 and 1980s.

'Politics Dressed Up as Science':
Political Participation, Environmental Democracy and the EPA

Introduction

One of the more enduring (if slightly disturbing) memories of my childhood was the sight of Norman Peak walking slowly up the road. It was a bemusing image of an old man moving in slow motion and all the more perplexing because it contrasted vividly with the energy being expended in a frenetic game of football. It was not that he was injured or anything like that. On the contrary, there was a sense of presence to his gait; upright, strong, even honourable. Yet, every five yards he would stop, bring out a handkerchief and cough, a gut-wrenchingly, deep cough. Every five yards. It would take something like an hour and a half to reach the pub; a walk that wouldn't even take us ten minutes. A lifetime of labour at the coalface had been reduced to a struggle for oxygen. My mother would casually explain that it was the 'dust'. To the medically trained, pneumoconiosis, a term I could spell as well as pronounce, which always seemed to impress the teachers. It seemed that the language of science, or at least that part of science which was relevant to mining, was a part of my everyday life.

From such a background it was not uncommon to see lichens. I can even recall going butterfly catching, an exercise in futility these days. Forget the red admirals or the peacocks, you'd be lucky to catch a cabbage white now. But lichens; lichens the size of footballs. Now that comes as a shock to the system. It occurred on my first walk up Ben Corr, Connemara. It was as if the environment had somehow overdosed on clean air. You see, that's the problem with pollution, it is insidious; it creeps up unbeknownst to you. Before you know it, the butterflies have gone. It is perhaps because of this that our everyday understanding of it is shaped by science, by its technically adept (and often unintelligible) prescriptions for what is happening. Science comes over as a neutral, objective, even heroic exercise. It seeks only what is best for humanity.

Perhaps it has something to do with my background but I am more sceptical. Mining had its own lexicon of technical jargon and, while the public may have an enduring image of a man like Norman Peak with a pick axe and a shovel, it was also an industry where the 'real power' lay in the hands of mining engineers. As in other industries, technological choices exist; it was just that mining engineers opted for those technologies which undermined the political power of those who worked at the coalface. As my father always insisted, 'it doesn't matter how they say it George, its politics, politics dressed up as science'. Science, engineering, politics; they're inseparable. There are political agendas in all walks of life and it is this, which forms a recurring theme of this chapter.

All too easily statements such as this can be misrepresented, dismissed out of hand as the thoughts of some 'green fundamentalist'; a tabloid-friendly phrase which carries an uncomfortable amount of ideological baggage and often elicits a political uppercut. It is important therefore to be clear. This chapter does not reject outright a role for science. Rather, it seeks to challenge and address its impact upon the democratic nature of environmental policy making. It is, therefore, critical of the current vogue, which accentuates the function of positivistic science, as if problems can be defined and responses assimilated in a purely rational manner. In its most simplistic form this approach tries to replicate the 'sterile conditions of the laboratory' in which science operates, conditions which underpin many of its assumptions and which are superimposed upon the political and social dimensions to everyday life. It is a paradigm which fails, not surprisingly, to accord any significance to the political dimension in policy making. And yet, as we shall see, contrary to its popular reception, and in key areas of policy, science remains uncertain, problematic and ambiguous.

The relationship between science, technology and environmental policy making has been an issue of increasing concern to political and social observers. Few, for example, would find difficulty in accepting the assertion that a scientific orthodoxy has enveloped environmental policy making, one all the more perplexing when we consider that most of the prominent issues in current environmental policy debates were raised by non-governmental organisations (NGOs) as long ago as the 1970s. As Grove-White observes, it certainly raises the intriguing question of how non-governmental organisations (very often poorly financed and organised) could have understood the issues in advance of 'official science'? (Grove-White, 1993). What is more, it is, in his opinion, striking that the analyses of organisations such as Friends of the Earth or Greenpeace have been vindicated on issues such as the impact of modern agriculture on vegetation and wildlife, the proliferation and damage of the modern car and the need for greater energy conservation to name but a few.

For authors such as Grove-White there is a need to caution against the prevailing orthodoxy which views science as a neutral and objective exercise. Indeed, he has argued that appeals to science to support one's position are inherently problematic and that 'uncertainty and indeterminacy in this area are not simply provisional; they are all but endemic'. What is often ignored or downplayed is that 'most scientists are responsible to institutions (industry, government, research councils)' and, as such, are 'necessarily as selective in their identification of "problems", and the relevant parameters which need to be considered, as are individuals or institutions using other modes of perception or understanding' (Grove-White, 1993, p. 22). As in so many other instances of life, the strength of science, its appeal to the rational, has also emerged as its achilles heel.

There is, for example, a growing body of opinion from political, cultural and environmental commentators which suggest that we have moved into a new era, a 'risk society' which has supplanted an industrial society, where science no longer possesses an overwhelming grip upon the public's imagination (see Beck, 1992, 1995). For Beck, it is a society increasingly critical of science, sceptical of its claim to be capable of organising, managing and controlling the escalating risk of modern society. Beck, for example, has argued that in a 'world-wide growth of large scale technological systems, the least likely event will occur in the long run. The technocracy of hazard squirms in the thumbscrews of the safety guarantees which it is forced to impose on itself ... the security system which anticipates social provision for the worst conceivable case, broke down with the advent of large scale nuclear, ecological and genetic hazards'. Whatismore, these hazards develop their own momentum and in doing so 'endanger the lives of everyone and stand in contradiction to the state's institutionalised pledges of safety and welfare' (Beck, 1995, pp 1–2).

If industrial society revolved around the distribution of goods, then risk society is about how we distribute the 'bads'. It is a transformation which begins where nature ends, as we shift our concern from what nature can do to us to what we have done to nature (Beck, 1998, p. 10). It abandons the assumption of marxism that industrial society is about divisions between classes. It is a post-modern condition because the hazards (chemical, ecological, genetic, nuclear) do not recognise inequality, national or geographical boundaries (see Lash *et al*, 1996. p. 2).

The idea of risk society might suggest a world in which serious hazards become ever more common, but this is not necessarily the case. In Giddens' view, for example, it means that society becomes increasingly preoccupied 'with the future (and also with safety), which generates risk'

(Giddens, 1998, p. 27). Crises such as that associated with BSE are not a matter of fate, but arise out of the complex interaction of decisions made by scientists, civil servants, regulatory agencies, industries and markets. In turn, a central paradox of this society is that 'these internal risks are generated by the processes of modernisation which try to control them' (Beck, 1998, p. 10).

In its latest variant, synonymous with the influence of Giddens (1998), it is alleged that society has moved from simple modernity to reflexive modernity. This is a move away from development understood in a unilinear fashion, that somehow modernisation follows a familiar and definite pattern, and acknowledges the emergence of a new order founded on the recognition of its own limitations (Giddens, 1998, p. 31). It is a society where challenges to the assumptions of epistemic elites (those who are privy to a specific form of knowledge or control its dissemination) are endemic.

There are potentially significant flaws in this approach. Wynne, for example, has suggested that, as it is currently formulated, it assumes that public mistrust only follows from open expert public dissent and contestation. However, 'it may well be that expert dissent is often only encouraged and sustained by the existence of a public backcloth of scepticism or alienation'. It should be acknowledged, therefore, that 'public alienation from and ambivalence towards expert institutions are not necessarily manifested in behaviour or overt commitments. No dissent does not mean it does not exist' (Wynne, 1996, p. 48).

These are not risks understood in the conventional sense, they are people-made hybrids, created, sustained and antagonised by the interaction between politics, science and public reactions. That is, society is at once both an issue and a problem (Beck, 1998, p. 11; Giddens, 1998). Although Wynne challenges the nature of scientific knowledge (see below) he also criticises Beck and Giddens for the failure to realise that, while the lay public may depend on the knowledge of experts, it does not mean that they do not mistrust them. Put simply, even in industrial society (and especially in coalmining communities) it was never the case that expert knowledge was simply accepted on trust. This clearly presents a problem for Beck and Giddens, who contend that there is an evolutionary movement toward risk society (see Lash *et al*, 1996, pp 6–9; Wynne, 1996, pp 43–83 and Welsh, 1993 & 1995). It is plausible to counter (if not entirely convincingly) that there will exist a period in which elements of both societies exist, before a fully risk condition emerges. It leads to increased pressure on scientific and political communities to control or minimise risk at a stage when difficulties appear to be imploding, leading to a marked preoccupation with 'manufactured uncertainty'.

This sort of argument sits comfortably within current debates about the declining role of the nation state within the new global economic order. Offe, for example, suggests that within Beck's work there is a drift away from the stage of national politics toward the public. Whereas society had previously been characterised by an increasingly interventionist state (at the expense of voluntary groups and associations) in risk society government can at best only offer points of direction, since its assumed responsibility is now deeply questioned. Precisely because the state can no longer control all the possible hazards they must at least devise mechanisms for the increased participation of active citizens. Thus, he states that:

> because there is no other force sufficiently 'sovereign' to impose a common good on them, they must control themselves, apply their capacity for practical judgement, and appeal to the cultural traditions of their form of life: they must substitute for a notoriously overburdened state power (Offe, 1992 p. 67).

Beck's work falls into a post-modern tradition which is searching for the conditions of a citizenship based upon active virtue. It has found accord among those who have dropped out of the materialist society, choosing to construct alternative informal economies where people exchange their skills rather than commodities. The link between these approaches and that of Beck is the desire for self-restraint, to limit or moderate the excesses of the capitalist economy and its detrimental impact upon our ecology. In Offe's opinion, it hinges on the metaphors of:

> brake and shackle – that is, on the intentional self-prevention of wrong moves ... be it organised science or state politics – to determine what the right moves are in theory or to execute them in practice is largely depleted (Offe, 1992, p. 68).

As O'Riordan's work on hazardous waste shows, however, even the terms risk and hazard are uncertain, imprecise and subject to 'inherent ambiguity'. Within positivistic science, which elevates the importance of the 'rational' and promulgates an ability to provide universal explanation, the inability to fully control or explain an event is attributed either to a lack of knowledge, ignorance or the competing views of experts. As a consequence of the need to exhort the rational, which forms the bedrock of the conventional approach of regulatory bodies, there is tendency to explain uncertainty or conflict simply in terms of 'technical imprecision'. Problems, which unexpectedly manifest themselves are, more often than not, 'concealed in the scientific language of manageable uncertainties' and are rarely ever aired in the public domain (O'Riordan, 1987, p. ix).

The current political furore surrounding genetically modified food is a case in point. With the political fallout of the BSE crisis still in the public consciousness, the hyenas in the media pack are encircling what they see as another carcass. Government officials and scientists appear before us to reassure public anxiety and to quell the tide of criticism over the rigour of the regulatory process. It is also a time when we see that most elusive of characters; the public face of multi-nationals. And it is a strange public face we see. Not on these occasions are we presented with managers, directors or public relations officers, but scientists. And not just any scientist. What we see on our screens is a clean cut, well dressed, youngish man (a man who could easily have a wife, two children, a mortgage and a dog). Quietly spoken, articulate and jargon free. The white coats, bald pate and spectacles have gone. The boffins of the rocket age who mesmerised us with a language which seemed to belong to another galaxy have disappeared. The scientists which the multi-nationals haul before us could quite easily perform the task of presenting the weather forecasts on Sky TV. The only tangible evidence that this is a scientist comes in the all important initials before the name, Dr Who(ever)?

Ultimately, risk becomes equated simply with the idea that we do not know, and it is at this point that 'penny begins to drop'. This is not the science we were taught in chemistry classes, where a combination of substances produces a known result. This is science in the unknown, where risk analysis becomes crucial. In other words, trust, such an important element to the acceptance of often unintelligible scientific justifications is at the point of being lost, with the potentially cataclysmic consequence that the edifice upon which modern science is constructed may collapse. It is an issue upon which O'Riordan's observations, written in the mid to late 1980s, are both illuminating and prescient:

> The eerie conclusion is that we are all busy enacting a public framework of regulation driven by certain fundamental tenets of rational knowledge – that *no one* believes in. Now there is always a certain tension between norms and actions. However the size and nature of the gap matters, it now seems so large as to approach collective hypocrisy – even institutionalised schizophrenia – on a grand scale. The public norms of rational control and decision in regulation seem to be little more than degenerate caricatures of reasoning (O'Riordan, 1987, p. ix).

A further source of difficulty presented in the prevailing use of science in environmental policy making, lies in the way in which expert and lay knowledge is differentially received by environmental agencies and government regulatory bodies. Here, Wynne's research into the relationship between scientists and hill sheep farmers in Cumbria, North England in the aftermath of the Chernobyl disaster, reveals tensions which surface when expert knowledge is challenged 'in the field'.

In the immediate aftermath of Chernobyl, the UK's Ministry of Agriculture stated unequivocally that it would not impact upon sheep farming in those upland areas most susceptible to radioactive fallout. It was a stance later revised, as the Ministry imposed a blanket ban on the movement and sale of sheep in the lake districts and the North of England, a potentially ruinous move for farmers. Amid confusion in the ensuing weeks, the Ministry decided to abandon this latter position, stating that a 'three week ban' would be sufficient given the scientific assumptions about the behaviour of radiocaesium in the environment. Scientists were convinced that, once it was washed off the vegetation and into the soil, the radiocaesium would be absorbed and 'locked away' (Wynne, 1996, p. 62). There *should* have be no further contamination, since the radioactive half-life of radiocaesium is twenty days and, therefore, all should have been well within three weeks, a view relayed with 'utter confidence unqualified by any hint of uncertainty' (Wynne, 1996, p. 3).

However, there was dismay among the farming community at the announcement in July of 1986 that the three week ban would be extended indefinitely. Contrary to reports from scientists, and the assurances of government ministers, the measured levels of radioactivity showed no signs of decreasing. It emerged from research over the following years that the 'three week assumption' had been based on research into alkaline clay soils, which were not normally associated with upland areas in Cumbria. In the acid peaty soils of the Cumbrian region, the radioactivity remained mobile and was, therefore, reintroduced into the food chain of the sheep (Wynne, 1996, p. 3). The research was also clouded by conflicting accounts of the 'background' levels of radiation prior to Chernobyl. This was due largely to the pollution attributed to the Sellafield (formerly Windscale) nuclear plant which has always attracted controversy (see Wynne, 1996, pp 64–6). Conspiratorial accounts abound. Both the coalmining unions and the peace movement insisted that the figures for both the economic costs of nuclear power and the pollution leaking from reactors was 'doctored' because of the need to sustain nuclear fuel for warheads.

Wynne's research highlights the fact that the scientists appeared unwilling to learn from specialist farming experience, a failing which contributed to the problematical conclusions of the research. Thus, while the research required measuring farms with different soils, management practices and climate, the scientists chose not to seek advice from local farmers who could have offered advice on where to make the relevant readings on the hillside. The assumptions and judgements in the field were uncertain and highly arbitrary. And yet, despite the flaws which later emerged, the scientific report obliterated any reference to 'uncertainty and open endedness by the time the knowledge returned to that same public as formal scientific knowledge in official statements' (Wynne, 1996, p. 66).

The problematical role of science not only presents a difficulty for government officials but also places environmental pressure groups in a quandary. On the one hand, they often wish to 'challenge the role of science', indicating not only its role in the degradation of the eco-system (CFC's, pesticides or genetic engineering) but also to interrogate the assumptions of neutrality, objectivity and truth. On the other hand, political pragmatism demands they recognise that, as currently structured, debates are almost entirely constructed in the language of science. As

Yearley notes, 'greens have good grounds for distrusting scientific authority but they have no other place to turn for universalistic procedures ... which leaves them open to the charge that they lack objectivity' (Yearley, 1993, p. 62).

This presents more of a difficulty for some than others. Campaigning groups, such as Greenpeace, are (or perhaps more accurately were) far more confrontational than conservation organisations and, therefore, tend to query more virulently the role of science (Dobson, 1990). Conservation groups on the other hand, draw their strength from a scientific or natural history background with a membership which typically involves scientists, ecologists and academics (Yearley, 1993). It is this group which is, more often than not, presented as the acceptable face of environmentalism. For others, there is always the dilemma that any attempt to question the role of science may well attract the tag of 'green fundamentalism'.

The tensions which reside in the relationship between science, environmental regulation and political protest have also been aggravated by the tendency among scientists to adopt the use of risk analysis. This was a contentious area of science which first surfaced in the public domain at the Sizewell B inquiry, where the methods and standards of safety which the UK Central Electricity Generating Board wished to adopt were investigated.

Amid the political turmoil surrounding nuclear power in the post-Chernobyl period, the Conservative Government in the UK chose to begin a major public inquiry into the proposed Pressurised Water Reactor (PWR) at Sizewell B (in Suffolk). The intention was that the inquiry would provide 'the fullest explanation and the fullest discussion' to date about nuclear power. It was anticipated that this would resolve the problems surrounding the politically controversial choice of the CEGB to use PWR technology, which had been used in the meltdown at Three Mile Island in the USA. The government was confident that the inquiry would vindicate the safety of nuclear power. It cost £25 million pounds (a very conservative estimate) and took two and half years to complete (see O'Riordan, et al. 1988). Without wishing to be alarmist, this is a case study which is all the more interesting (and worrying) because it involves many of the issues which have surfaced in the political furore which now surrounds experiments into genetically modified organisms. In the language of the word processor, simply highlight nuclear and replace with genetically modified crops or food.

To a large extent, many of the difficulties in the areas of nuclear power or GMOs can be located in the overwhelming scientific complexity of the debates. It is a problem which has not eased with the passage of time. Compare, for example, the request from Mr Justice O'Sullivan in the recent case brought by Genetic Concern on whether they could agree to a simplified guide to genetic engineering to which he could refer, with those of the Inspector into the proposal to build an advanced gas cool reactor in the UK in 1971:

> Nuclear risk is a subject so sophisticated that we, like the objectors, have no reliable grounds for framing an independent opinion, and we have been obliged to accept both the assurances given by the experts called by the generating board and those implicit in government policy ... (cited originally in O'Riordan *et al*, 1988).

Almost three decades on and little appears to have changed.

A large part of the Sizewell B inquiry (which ran for 340 days) centred on the use of risk analysis in the construction of safety measures designed to prevent the meltdowns which occurred at Three Mile Island and Chernobyl (see O'Riordan, Kemp and Purdue, 1988). It is a commonplace, both in the published literature, arguments and public statements of the EPA and multi-nationals, to use terms such as 'acceptable forms' of pollution and acceptable risk. They are phrases with a 'touch-feely' sense of reassurance and yet they are, as O'Riordan *et al* observe, essentially meaningless

(O'Riordan *et al*, 1988, p. 192). Despite this, the term acceptable risk (and its familiar cousin, tolerable risk) has assumed a quasi-religious hold on environmental policy makers, a tendency discernible both in the increased use of risk analysis and the burgeoning field of research which seeks to justify and expand its use.

For many within the scientific community probabilistic risk assessment is the key to good engineering reliability analysis. To the more sceptical it is misleading, seeking simply to mesmerise decision makers into judgements that place a bias on technical factors rather than equally if not more relevant social, political, economic or ethical issues (O'Riordan, 1988, p. 193).

At the Sizewell B inquiry there was conflicting evidence about what represented an acceptable level of risk and the probability of a major nuclear catastrophe. Indeed, as the inquiry lengthened and its focus narrowed, it revealed a subtle, but nonetheless crucial shift in the way in which the Nuclear Installations Inspectorate (NII – the regulator) approached the issue of safety probability.

Prior to the Sizewell B inquiry, the NII had adopted what was termed an 'effective barrier concept', a set of engineering provisions designed to prevent releases or reduce the size of releases. It is not a physical structure as such, rather a group of safety provisions which insist upon strict standards and the inclusion of a diversity of effective barriers. Clearly, the more barriers, the less likelihood of rogue releases. The drawback was that the more barriers, the greater the cost.

At the inquiry it began to emerge that the NII no longer deemed the 'effective barriers' philosophy necessary, preferring to concentrate on the 'bounding of fault sequences', a procedure adopted in the USA. Crucially, it emerged that the driving force behind this change had come not from within the NII, but the engineering fraternity in the CEGB (O'Riordan, 1988, p. 217). In other words, it was the licensee which had requested change, a shift in safety philosophy borne not out of any proven superiority of a new method, but because experts in the CEGB favoured an approach which differed from experts in the NII. It is not, and let us be clear on this matter, about brown envelopes or shady deals. It is an area of science understood by few, and those in the 'know' recognise and accept this uncertainty (even though they may be reluctant to admit as much in a public environment).

More importantly, the shift in safety philosophy had been undertaken not because of any improved scientific proposition but because of a combination of the political interplay between the regulator and the licensee and the internal dynamics of the engineering fraternity in the CEGB. This is by no means an exceptional case. In such organisations it is rare for all to be in accord; dissension and disagreement are almost endemic to such working environments.

The nature of uncertainty and scientific fallibility was a recurring theme at the Sizewell B inquiry, where debates about the reliability of nuclear technology dominated the agenda. In an attempt to establish a common ground, the inspector asked two eminent scientists to examine issues surrounding technological reliability and risk analysis with the stated aim of discovering whether he could go the Secretary of State and say; 'you may be reasonably assured that now it is in the hands of the statutory licensing authority and the public may rest content that safety will be adequately protected' (O'Riordan *et al*, 1988, p. 229).

Dr Gittus, one of the experts consulted, suggested that risk should be able to withstand three tests; importance, consistency and completeness. What mattered was the character of the danger, the threat for the public and the magnitudes of likelihood and frequency. However, as O'Riordan observes, this dependency on a positivistic approach to science encounters difficulties in areas such as risk analysis where these conditions do not always apply

(O'Riordan, 1988, p. 229). In particular, he has suggested that both policy and risk analysis neglect the importance of implementation, which is much more than 'mechanical or local enactment'. Deviations or implementation deficits are often related to local factors which are generally underestimated or downplayed within risk analysis, which tends to assume a rational or methodical transfer (and acceptance) of knowledge (O'Riordan *et al*, 1988, pp 13–14). There is no room here for competing views, conflicts of opinion or differences in how the changes are received. In short, it neglects the institutional context in which there may be 'different definitions of the decision problems, different perceptions of what the primary risk generating problem is, and different kinds of relevant experience and expertise' (O'Riordan *et al*, 1988, p. 14)

Even if we do not subscribe to the view that a risk society has developed, it is still plausible to suggest that a catch 22 situation seems to have emerged, where regulatory bodies strive to achieve credibility through the dissemination of complex scientific knowledge in a period in which uncertainties appear to have escalated. The outcome, as O'Riordan observes, is that:

> as expertise is increasingly used to justify regulatory decisions, it is increasingly challenged, leading to public examination of the scientific process. This is found to fall short of (unrealistic) formal, rule bound images of scientific practice, leading to further deterioration of regulatory credibility (O'Riordan *et al*, 1988, p. 14).

Probabilistic risk assessment, acceptable risk, the likelihood of an event occurring in the range from 10^{-2} to 10^{-6}, this is the mystical terminology of the science of risk. An obscure area of science dominated by mathematical modelling which, in a nutshell, does not guarantee that an even will *not* occur, just that it is unlikely to take place. To put the problems in this area into perspective, the Sizewell B inquiry took 150 days (as well as a huge number of written submissions from experts) to discuss the issue of safety and the role of risk analysis. Little in the way of a definitive conclusion was drawn.

Or, put it another way. One of the key tenets of risk analysis is the use of comparison as part of a process of fine tuning. Yet, with highly complex technologies such as nuclear power or GMOs, we simply do not know. We have no experience to draw upon, an omission dramatically revealed at the Sizewell B inquiry where the project director of the pressurised water reactor at Three Mile Island was asked if he 'had any comparable experience?' His response, which provoked laughter in the audience, was 'No. (not) until after the accident' (O'Riordan *et al*, 1988, p. 182).

As Maguire seems only too well aware, 'too often, we hear the experts calling, each to each. We need to hear what the expert is not telling us, has indeed very likely not told her or himself'. What we should demand to hear more are often 'those things too silly, too wild, too embarrassing for normal speech'. In other words, science operates within a language where the 'howls are restrain'd by decorum' (Maguire, 1996, p. 170).

We are now witnessing a tension which resides in the relationship between science and politics. It is a difficulty which manifests itself increasingly in public dissatisfaction with scientific judgements on environmental matters. It is not simply a problem for science, a failing if you like to explain itself more clearly. No. The roots to this problem lie in the manner in which science has been used in environmental regulation; to obfuscate rather than illuminate in situations of uncertainty. A failing to allow greater participation and transparency has been compounded by the way in which regulatory agencies have turned to science to justify decisions which have been removed from the public sphere. As the public's anxiety has grown, and political protest has surfaced, governments and regulatory agencies have chosen to 'side-step' the issue(s), choosing to reassure through the media. Calls for greater participation or more stringent regulation of GM technology are seen as the cries of the heretic, anarchists or worse, green fundamentalists, 'out of touch' with the imperative of sustaining a healthy economy.

These are charges common to debates in Dáil Eireann where, often as a last resort, politicians find the temptation to present a stark juxtaposition between the economy (jobs) and the environment irresistible. Although by no means alone in being susceptible to this line of thought, Brendan Howlin, a former Minister of the Environment, captured the essence to this argument when he suggested that:

> the Green flat earthism we have heard twice today in this house does no service to the cause of trying to create jobs in this country. The 300,000 people out there who are looking for jobs will thank us in no small way when we talk in jocose or simple terms of technology being bad in the most basic and luddite of terms which is patent and absurd nonsense (Howlin, *Dáil Debates*, 1992, vol. 418, p. 957).

It should come as little surprise then that decisions once adjudicated upon by civil servants, public officials and regulatory agencies have now become a source of intense public debate and controversy. In resolving the type of political protest likely to ensue on subjects such as landfills, waste incinerators or the construction of new roads, public hearings, in whatever guise, have become a mainstay of measures designed to ease fears among environmentalists that regulation is open and transparent and that decisions are not imposed without informed public debate and consultation. As in the wider political debate which surrounds environmental policy, science has also played a pivotal role in shaping the nature of oral hearings, and it is to this that we now turn.

Oral Hearings, Political Protest and the EPA

The EPA Act has implemented a structure in which oral hearings perform an advisory function, a move which sits comfortably within the Irish administrative tradition which prefers an adversarial format as a means of gaining evidence. In such circumstances, the inspector's role is not to produce a decision, rather it is to collate information and report back to the EPA. It is the inspector's task to define what is useful and what is not. In the final report he or she may make recommendations, but the decision ultimately lies in the hands of the EPA (or the Minister).

In contrast to judicial procedures, proceedings at oral hearings tend to be less formal, less concerned with protecting individual rights and more with ensuring efficiency, flexibility and administrative discretion. Cross-examination is not a right, nor indeed is attendance at an inquiry, except for those who show a direct interest (see O'Riordan *et al*, 1988). Hearings tend to be loosely structured, open forums where members of the public hear government proposals and have the opportunity to respond, typically in a format set by the agency.

One of the more common objections to oral hearings is that they rarely provide citizens with an opportunity to participate on an equal footing, since the agency defines the agenda and establishes the format. In addition, they allow only a limited amount of time for citizens to grasp technical issues and offer a substantive element to discussions. These are difficulties compounded by fact that the EPA Act stipulates that where an objection has been lodged, the agency shall have *absolute discretion* to hold an oral hearing which shall be conducted by a person appointed by the agency (EPA Act, 1992, p. 64). It is a legislative approach strong on procedure, stipulating time periods for the submission of written evidence, the decision to hold an oral hearing and the notification of decisions taken by the agency. However, by far the most glaring omission is any clause which may hold the agency to account. In short, the notification of a decision remains a far cry from an explanation of 'how and why' a particular decision was taken. It remains one of the more unedifying paradoxes of the EPA's legislative framework that on the one hand, it is charged with disseminating environmental information and yet, on the other hand, it is denied (or fails to provide) a forum in which this information may be engaged seriously.

There is also a subtle, but nonetheless important, interplay between the procedures at oral hearings and the role of science, one which drew little attention during the formation of the EPA Act, but which has become increasingly controversial. Influenced by changes in planning legislation in the late 1980s and early 1990s, the EPA Act adopted a 'one shot' appeal system and imposed a time limit in which an objection from a third party could be lodged (21 days). Both of these moves were motivated by a desire to streamline the appeals process, curtail 'vexatious and frivolous' objections and establish what constitutes a 'valid objection'.

A valid objection? You may be forgiven for thinking that a well informed, carefully selected argument would constitute a valid objection. This was certainly the view taken in the decision in R. v. Inspectorate of Pollution ex parte Greenpeace Ltd (no.2) where it was stated that:

> it seems to me that if I were to deny standing to Greenpeace, those it represents might not have an effective way to bring the issues before the court. There would have to be an application either by an individual employee of BNFL or a near neighbour. In this case it is unlikely that either would command the expertise which is at the disposal of Greenpeace. Consequently a less well-informed challenge might be mounted which would stretch the court's resource and which would not afford the court the assistance it requires in order to do justice between the parties (cited in Simmons, 1997, p. 100).

In some quarters, it was anticipated that oral hearings would provide the forum in which objectors could mount a carefully selected, focused, relevant and well argued challenge. However, the EPA legislation has little sympathy with such sentiments. This is most evident in the enunciation of the terms for a valid objection, where the EPA's guidelines stipulate that a 'valid objection' requires not only that all of the 'necessary documentation' is supplied in the initial application, but prevents 'further submissions' being made (EPA, 1994, p. 11). The timing of the objection becomes crucial, and acts as a further constraint upon the ability of pressure groups to construct a strong case.

Place yourself in a situation where a decision has been taken to locate a waste incinerator or a landfill near your home. You have 21 days in which to set up an opposition group, contact any favourably disposed leading scientists or legal experts, discuss the approach and make an objection. There is no 'second chance' to add information at a later stage, should it become available. And all in 21 days. Put simply, in an era in which science (and in particular the more disputed areas of science) have become the focal point of objections, third party objectors are finding it ever more difficult to construct a well-argued case given the time constraints which now exist.

It should be noted that this is not a conundrum faced by licensees (business) since the agency has declared that it is willing to go 'as far as possible to assist industry' and to 'provide pre-application clarification facilities where necessary' (EPA, 1994, p. 5; Taylor, 1998b). There can be little doubt that, at times, it appears that the both the regulatory framework and the agency positively balk at the political prospects which accompany any increase in participation.

Political Protest, Waste Incineration and the Oral Hearing in Clare

The oral hearing on the proposed incinerator for Syntex in Clare highlighted many of the difficulties third party objectors encounter in these forums as they are currently organised by the EPA. There were three issues which dominated the agenda; the question of incineration as opposed to waste minimisation, the scientific debate on the effects of dioxin and finally, the legitimacy of the oral hearing itself (Taylor, 1998b).

The central focus of the scientific debate at the hearing in Clare centred on dioxin release. In its quest to delay the licence application, the Clare Alliance Against Incineration (CAAI) argued that any prudent judgement should await the new guidelines on toxic emissions to emerge from the European Union and the US EPA. However, it became clear in the course of the hearing that there were conflicting expert opinions with no single, unified scientific position on dioxins (Taylor, 1998b). It clearly revealed the problems of adjudicating on complex scientific areas, where appeals made to 'objective' scientific evidence to establish the validity of a given position are problematic. However, this was the terrain favoured by the hearing, where the parameters were drawn along the lines which recognised the need to establish 'scientific truth' or economic viability, rather than any discussion of principles of environmental justice.

However, of far more import to third party objectors was that the onus was upon the environmental lobby to prove that the incinerator would be environmentally damaging, rather than the polluter to justify that it is environmentally benign (Taylor, 1998b). It is a subtle, but nonetheless crucial distinction which favours business (or the licensee), and emerged in the case between Genetic Concern and Monsanto over genetically modified crop trials. As a forum for political debate, oral hearings also favour the industrial lobby in the way in which they shift the terms of debate away from environmental *protection* towards environmental *management*. They function not as an arena in which to discuss the prevention of pollution, but rather to establish criteria for what is an *acceptable level* of pollution, usually within a defined set of parameters which do not *impede* economic growth. This is a problem compounded by the lack of resources often available to third party objectors who are required to confront multi-national companies which possess vast resources and access to trained lobbyists. Constructing a scientifically rigorous case may not only be time consuming but expensive, which only serves to favour business.

The final objection raised by the environmental lobby toward the oral hearing in Clare was on the procedure adopted, with a single person acting as 'chairman, recorder of all evidence and indeed assessor of the evidence to be given at the four day hearing' (Taylor, 1998b). As the spokesperson for Greenpeace Ireland remarked, that a decision on the proposed incinerator should end in an:

> oral hearing, presided over by one man with a notebook, is an insult to the people of Clare and testimony to the major inadequacies of the EPA Act (Greenpeace Ireland, *Clare Champion*, 4 Oct, 1996).

The hearing in Clare also revealed the chasm which currently exists between the EPA and the environmental lobby with regard to the function of oral hearings. As far as the EPA is concerned, oral hearings are simply one of a number of mechanisms by which it garners environmental information. They do not necessarily take place because of the weight of public opinion or the number of objections. Indeed, as far as the EPA is concerned, it matters not whether a project or licence has 5000 objectors or one. This is not meant to be a flippant remark, it merely reveals the sentiment which resides in the EPA's position; it is the 'quality of the objection' or, perhaps more accurately, the extent to which the objection is favourably received by an epistemic elite privy to an exclusive interpretation of the balance between BAT and NEEC (Taylor, 1998b). In other words, objections which are likely to succeed are those which the EPA will find favourable. It hardly requires a great leap of the imagination to see that Nimbyism is hardly likely to strike a chord within the EPA.

Agencies such as the EPA also fall prey to what Maguire has termed the 'Could Try Harder Syndrome'. They remain ossified in a view of the world where we appear to be moving in the right direction, that we should concede that 'Rome wasn't built in a day, and that what is required is to sustain the quest for more information' (Maguire, 1996, p. 171).

The position of the environmental lobby on oral hearings could hardly be more different, where they are perceived as an opportunity to challenge or *prevent* polluting activities, rather than simply *manage* them. However, to its dismay the parameters drawn rarely extend to the possibility of removing licences, rather the preference is to attach more conditions. As a forum to shape policy they are, therefore, crucially limited, even emasculated by an operating philosophy which underpins BATNEEC and accords greater emphasis to making concessions and organising compromises. Emission levels, audits and economic viability. These are the values with which the EPA is familiar. It is not about positing 'absolutes', rejecting licences outright or adjudicating upon complex political debates about environmental justice. In a nutshell, oral hearings are an institutional manifestation of the EPA's operating philosophy: environmental management.

What is to Be Done?

Yearley has suggested that since nature and wildlife are not capable of 'speaking for themselves', there is a need for a stand-in. However, as he points out, while it is only scientific knowledge which can perform this role in modern society, it is a role with which it is uncomfortable (Yearley, 1993). While I have some sympathy with Yearley's observations, particularly with the difficulty environmental groups have with science, I am not altogether convinced that the principal source of our difficulties lies with science *per se*. It is not a matter of more lay knowledge or less science, but about creating a democratic framework capable of enhancing participation and reconciling environmental conflict. These are complicated (and controversial) issues, and before we can address the potential for Irish institutions to construct a 'successful' ecological policy, we need to detail briefly some of the more prominent impediments which lie in the path of a democratically informed environmental policy. Let us place these issues in some sort of context. If we were to follow O'Riordan it could involve adopting the precautionary principle where thresholds of renewal are in doubt. Or, redirect profits in order to secure new resources or ensure an orderly succession of alternatives. It may necessitate the protection of critical life support processes (species and habitats); to ensure minimum standards of living for all people and to guarantee basic freedoms such as expression, education and liberty. In anybody's language, a formidable challenge (O'Riordan, 1996, p. 140).

It would be extremely misleading to begin any discussion of the principal obstacles to environmental policy without addressing the relationship between the economy and the environment. Here, there are at least two key themes which demand consideration. First, the issue of whether economic growth can be 'sustainable' (at or even near its present limits) and, secondly, whether it is possible for 'nation states' to create and sustain a policy stance in isolation of the pressures which arise from global business decisions.

There is no shortage of publications from governmental and non-governmental sources which purport to address the issue of sustainable development. From the Bruntland Report to the conference at Rio de Janeiro, we have been inundated with 'recommendations', discussion documents and frameworks for the future. Indeed, it is almost impossible not to be impressed by the sophistication of the political rhetoric which often accompanies these proposals. Substantive change, however, has proven far more elusive. This has not been helped by the fact that the term 'sustainable development' has been used with such gay abandon as to be almost meaningless. There is, as O'Riordan notes, little in the way of a coherent picture of what a 'sustainable society would look like ... what we tend to talk about in practice is a society that is unsustainable' (O'Riordan, 1996, p. 140). These dilemmas have only been complicated further by the acceleration toward a global economic order. It is not just that financial speculators can wreak havoc upon national economies, itself a worrying development. Rather, it is the pressure imposed on national decision making by the dynamics of the global economic order which poses the larger threat. We need go no further than Seagate's decision to pull-out of Clonmel, or Apple's recent decision to withdraw production of the iMac from Cork, to recognise the 'complexity' of the decision making environment in which multi-nationals operate and the subsequent impact they can have upon local communities.

It is imperative, therefore, that as a collective community we recognise that rigid environmental legislation, or even legislation more stringent than that adopted by our perceived competitors, may serve to jeopardise potential investment. This is not a new phenomenon, particularly for economies such as Ireland, which have been dependent upon international investment, it's just that it appears to have accelerated in the 1990s. It is at this point that we can also appreciate more fully the interface between the environment, science, the economy and politics. These are real issues, political issues which impact upon national and local communities and frame the nature of environmental policy making. To think otherwise would be naive in the extreme. In short, if we choose a path which maintains stringent regulation we need to acknowledge the political, economic and social consequences for the community at large.

Competition at a global level also raises the spectre of nations creating pollution havens in order to 'attract' the potential multi-national clients, driven by an inexorable quest for lower production costs. Issues such as this reinforce the cleavage which exist between the aspirational and the pragmatic, the possible and the idealistic. Yet, it is important that we avoid at all costs becoming mired in sterile economic arguments which shift in a subtle manner from being about the economy *and* the environment to the economy *or* the environment. They are certainly difficulties not alleviated by the timescale in which environmental policies tend to be formulated and implemented. Pollution problems very often develop only slowly; the loss of individual plant species, insects or birds may evolve only after a prolonged period of time, presenting a dilemma for those who wish to bring these matters to the attention of a public weaned increasingly upon the importance of the 'immediate'. It is hardly surprising, therefore, that few politicians see 'votes' in these matters, they are issues which do not synchronise comfortably with the political cycle of the Dáil.

Although the nature of environmental democracy is an extremely contentious area of 'green thinking', there appears to be a consensus forming around the need to establish a stronger link between environmental issues and enhanced political participation (Lafferty and Meadowcroft, 1996; Dobson, 1996). However, while this has been an area which has attracted much attention from political theorists (Goodin, 1992; Dobson, 1996 and Saward, 1996 to name but a few) it has been relatively neglected in public policy circles, where the tendency has been to examine the more pragmatic avenue of new policy instruments (green taxes etc.). A search for the attainable or the probable, rather than the possible.

For some this is by no means as clear-cut as presented here. Lafferty and Meadowcroft, for example, observe that it may be that acute environmental problems are 'more readily (or perhaps only) amenable to an authoritarian solution

(Lafferty and Meadowcroft, 1996, p. 3). It is a theme to which Saward has warmed to, suggesting that there are no good reasons why green assumptions about the compatibility of democracy and ecology should be accepted. Indeed, there is a distinct possibility that when the values of greens and democrats conflict, it may well be the case that democratic values are jettisoned (Saward, 1996). Saward's contention is not that green arguments cannot be included in democratic systems rather, that where:

> green outcomes take precedence over strictly democratic outcomes, it ought to be recognised and acknowledged that democracy is being diluted. There is no necessary prescription that democracy must win – or win fully – when principles conflict in practice, but at the very least the external values employed ought to be defended explicitly (Saward, 1996, p. 93).

These remain crucial issues for many greens, who often espouse the need to improve democratic channels in decision making and exhort the need for participation (see B Doherty and M De Geus, 1996).

There are few who would doubt that such difficulties have been influenced by a number of separate, but inextricably linked trends which have questioned previous forms of state-led environmental regulation. Influenced by the British preference for pragmatism the EU has, for example, eschewed the grand plans of the 1980s (supranational policies which sought to usurp variations at the national level) and has concentrated upon how policy can be implemented and monitored. This has been manifest in the combination of the IPPC regime and the adoption of the 'polluter pays' principle, both of which have their origins in the difficulties the EU experienced in 'delivering' EU wide agreements on policy. While these measures accelerated after the demise of Mrs Thatcher's Conservative government, the administrations which she led left a powerful legacy which sought to challenge the role of the state in environmental regulation.

The rise of the 'polluter pays' principle, for example, has clearly been influenced by the New Right's critique of state intervention and the perceived need to expand the role of market instruments to reduce pollution. The New Right refers to a loosely constructed coalition of interests between liberal economists and the socially conservative politicians synonymous with the political administrations of Thatcher, Reagan and Kohl. Put simply, the quest of these administrations was to 'roll back the frontiers' of the state, release the 'invigorating' forces of the free market and gain the benefits of competition. The principal sources of inspiration lay in an amalgam of ideas gleaned from the liberal political philosophy of those such as Hayek (1944) *The Road to Serfdom* and the free market economic thinking of Adam Smith (1776), Milton Friedman (1962) and Buchanan and Tollock (1975). As a political tradition it rejected policies organised around rigidly defined standards imposed by regulatory bodies in favour of free market solutions which accord the consumer sovereign status. Its principal source of angst lay in the role of the state, the inefficiency of bureaucracy and the need to reorganise the public sector.

This concern with bureaucratic failure can also be found in the work of those such as Janicke, where problems with environmental regulation over the last three decades are attributed to 'state failure'. However, he is at pains to stress that this cannot be reduced to regulatory capture (a cosy relationship between the regulator and the regulated), rather it involves the need to explain the drift in policy which has occurred. This drift in policy, according to Janicke, is one of the major obstacles to the formation and implementation of a successful ecological policy and has its roots in the nature of relations between bureaucracies, politicians and industry which tend to adopt routinised solutions less conducive to innovation (see Eckersly, 1995, pp 12–25).

Given the general tenor of Janicke's critique of state intervention it is hardly surprising that his recommendations should seek to achieve a reduced role for the state and accord a corresponding expansion to active citizens. If a transformation in technological efficiency and a radical change in production structures is to be exacted then it is, in Janicke's opinion, possible only through the promotion of participation and decentralisation (Janicke, 1996, p. 71). The evidence he elicits to defend his position is drawn from a series of comparative studies and suggests that if a successful ecological policy is to emerge, and a policy stasis avoided, we need to construct a vibrant oppositional sphere in civil society to force the pace of change (Janicke, 1996). It is a position which reflects a disillusionment with state-led regulation and the institutions of representative democracy (parliamentary codes of practice). Consequently, for Janicke, the line of resistance to the environmental excesses of business, and the construction of a successful ecological policy, have their origins in improved 'capacity building', enhanced social learning and improvements in the constitutional rights of oppositional forces (pressure groups) in society. In other words, the strength of Janicke's position can be found in the 'establishment of a well functioning network of green interest organisations, industry and environmental administrations' (Janicke, 1996, p. 78).

Appealing though initially this seems, it is a view which imposes an undue level of co-ordination and coherence upon the role of the environmental lobby. It certainly appears to downplay the fact that most environmental protests tend to be short-lived, localised, and even accidental. Neither does it take cognisance of the disparate and often conflicting agendas which pervade the environmental lobby.

Janicke certainly treads a well-worn path taken by contemporary political scientists and environmental observers who, informed by the collapse of the east European socialist states, disillusioned by the (perceived) failures of social democratic state intervention in the West, and abhorred by the excesses of the Thatcherite era, now implore the need to 'reconstruct' the citizen as the active bearer of a new politics. It is an approach which seeks to break out of the impasse posed by the old dichotomies; capitalism and socialism, state and the market, capital and labour, all of which are deemed inappropriate to the modern era. It is the politics of the third way, bereft of issues such as power, struggle or exploitation. It focuses upon negotiation and consultation, of informed and passionate debate. There are no zero-sum games here, and very few losers. Welcome to the age of conviviality.

If I am sceptical of Janicke's views on the positive gains to be made from the disparate 'oppositional forces of civil society', I am more persuaded by his views on policy integration. On this matter he has correctly observed that environmental policy generally starts with the establishment of a new agency or bureaucracy. However, such institutional differentiation has little effect, if the capacity for intra and inter-policy integration remains low. For Janicke, overcoming the incremental and isolated role of environmental administrations within government emerges as an important prerequisite of a successful environmental policy.

A further theme important to our discussion of environmental democracy, one which has figured prominently throughout this book, has been the power of influential pressure groups. It is a common objection to the atomised character of modern pluralist states to argue that they have become over burdened by the complicated and often conflicting agendas of pressure groups. A sclerosis in policy emerges, a consequence of the myriad of demands from pressure groups.

To its critics, pluralism (at least initially) ignored the fact that groups possess vastly disproportionate resources and that the 'ear of government' is more open to the arguments of business, upon which the health of the economy (and a government's future) ultimately depends. To others, such as Schmitter and Lemhbruch (1979) the emergence of organised bargaining in the 1970s between trade unions, government and business (corporatism) significantly undermined the pluralist perception of government/pressure group relations as being arranged around open dialogue.

In moves designed to bring stability to a political sphere shaken by two oil crises and rampant inflation, it was argued that modern democracy had altered irrevocably. It was no longer about a 'plurality' of interactions between government and pressure groups, but about an organised relationship of bargaining between the state, trade unions and business. Predicated on state intervention it 'fell out of fashion' in the 1980s, as the high ground of politics in western Europe and the USA was dominated by the agenda of the New right.

In an attempt to seek an institutional set-up conducive to the environmentalist demand for participation authors such as Lafferty and Meadowcroft have resurrected corporatism eco-style, where the emphasis is upon the co-operative, integrated and inclusive dynamic of these relationships. While reticent about using the term corporatism, the essential characteristics displayed in their term, 'co-operative management regimes', appear familiar: a 'type of formation which involves a number of social partners in a collaborative attempt to resolve specific environmental matters' (Lafferty and Meadowcroft, 1996, p. 257). Thus, for example, success ultimately hinges upon the ability to construct a discursive consensus. In other words, groups (both environmental and non-environmental) need to interact with other social organisations and become engaged in a (positive) exchange of ideas and arguments. It is an institutional set-up based upon negotiation and compromise where, crucially, agreements require that all parties recognise their legitimate status. By the mid 1980s, with the New Right's political onslaught in full throttle, and burdened by conflicting interpretations of what corporatism actually meant, the concept fell out of fashion. As with 'flares' they were seen as an aberration of the 1970s. The widely perceived view was that it was an unduly rigid form of decision making, ill-suited to the demands of the new global era (see Lash and Urry, 1987; Streeck, 1992 and Gobeyn, 1993. For critical comments see Taylor, 1996).

A second dominant theme is that if stability is to be secured, and policy continuity sustained, then each participant must be able to deliver on their 'part of the bargain' (Lafferty and Meadowcroft, 1996, p. 257). As Lafferty and Meadowcroft point out, an 'emerging co-operative management regime', demands that participants must to 'some extent transcend their status as mere advocacy groups for particular constituencies' in order to reinforce the legitimacy and credibility of the policies agreed (Lafferty and Meadowcroft, 1996, p. 257).

There are at least three important sources of tension within this approach which demand consideration. First, the groups, which are usually 'included' within the process of negotiation, are not necessarily organised along democratic lines. Second, the process of bargaining is removed from the traditional arenas of parliamentary democracy (the houses of the Oireachtas) and is not therefore subject to the 'normal' democratic mandate. They involve negotiations which take place largely outside of the control and scrutiny of elected representatives, a process which ultimately reduces the role of citizen as voter and may subsequently impact negatively upon the active citizen. Thirdly, such methods of policy formation (which involve negotiation, deliberation, consultation and compromise) can often be extensive and convoluted. To critics they represent an archaic form of 'politics', far too rigid for the demands of the new flexible, global era. And yet, while many of its west European counterparts have abandoned these institutional arrangements, Ireland has persisted (successfully) with them since 1987 (Taylor, 1996).

As O'Donnell observes, one of the key dimensions to this process of 'partnership' has been the detailed, shared analysis of economic and social problems (O'Donnell, 1998).

The virtues of this approach have been reinforced by the NESC (1996) report, *Strategy into the 21st Century*, where it was argued that the Partnership process involves different participants in a range of items from national economic policy to local development.

There were two key elements highlighted by NESC which have been significant in the transformation of the Irish economy: functional inter-dependence (bargaining and deal making) and solidarity (inclusiveness and participation) (O'Donnell, 1998, pp 101–102). However, as O'Donnell has suggested, there is a third dimension; deliberation. While bargaining and negotiation may distinguish Partnership from its pluralist counterparts, where consultation is more prominent, it does not entirely capture the essence to this approach. For O'Donnell this lies above all else in a process of deliberation which has the capacity to shape and reshape a participant's understanding, identity and preferences. Implicit to this approach is the idea that deliberation involves a shared understanding which coalesces around 'problem solving'. Partners tend not to debate their ultimate 'social visions', and consequently consensus is not a pre-condition for Partnership, but an outcome (O'Donnell, 1998, pp 101–102).

These issues have become an increasingly familiar part of Irish political debate, as academics and politicians grapple with the source(s) of Ireland's recent economic success. Certainly, the pessimism which had prevailed among political historians at the beginning of the decade (J J Lee, 1992) has been replaced by an overwhelming political ebullience encapsulated, if not altogether explained, in the concept of our times; the Celtic Tiger. There are those who may dispute individual features of this phrase, but few would deny that it seems to possess that most ethereal of qualities, the ability to act as a prism through which our disparate, individual experiences of the contemporary Irish polity are refracted and made intelligible (Taylor, 2001).

On the one hand, it seems capable of pervading the currency of the mundane aspects of everyday life, confirming our intuitive sense of change; from the transformation of the workplace experience to conversations which express dismay at either escalating house prices or the time we have spent in Dublin's gridlock. On the other hand, it has found considerable favour in political and media circles, where it has rendered intelligible the almost incomprehensible: a period of unprecedented economic growth unimaginable a decade ago (Taylor, 2001).

For both political scientists and neo-liberal economists, such a reversal in the fortunes of the Irish polity is all the more puzzling because it has occurred against the backdrop of a period dominated by a succession of national level wage bargaining agreements (Partnership). For those economists of a more conservative persuasion these agreements are an anathema to a healthy economy, since they intervene in the natural workings of the free market. And yet, perversely, while those of the conservative right are keen to distance themselves from this hybrid, or at least remain less prominent in the letters pages of the *Irish Times*, there are those of a social democratic hue who see it as little more than Thatcherism in wolf's clothing. If we are to locate the source of this confusion it is critical that we recognise from the outset that the term does not expresses anything uniquely Irish or indeed Celtic. Rather, its allure rests on something it is definitely not. It is definitely not Thatcherism. The ease with which the Celtic Tiger metaphor has been so widely embraced lies in the fact that in some intangible fashion it confers upon an economic strategy a social democratic ethos (or at least its proponents hope it does) which is still capable of securing a place at the table of the global economic leaders without enduring the divisions which have riven British society. Although political debate in media circles has often been reduced to little more than discussions about the benefits of reductions in taxation or increases in welfare provision, the central issues remain far more complex and deep-rooted (Taylor, 2001).

There can be little doubt that the Irish State has occupied itself with an altogether more important political project, one designed to engineer a new form of governance capable of reconciling the demand for the level of economic growth associated with other major European countries with the need to ensure at least a modicum of social inclusion. Thus, in contrast the Thatcherite project in the UK which disassociated itself from discussions with the trade union movement, Ireland has been proactive (and at least partially successful) in pursuing a programme of economic restructuring which has included the unions.[25] That is, such intermediation has contributed positively to the creation of a set of political and economic conditions favourable to management. In other words, this corporatist strategy has allowed a restructuring of the *supply* side of the economy (the introduction of new technology and new work practices) which is in line with the prime objective of remaining competitive in the new global economic order.

The tension which resides in this approach (and is the principal source of confusion) is that a set of national level negotiated agreements contain elements of a neo-liberal economic and political project. There are areas of public policy, for example, which bear the fingerprints of a neo-liberal approach (that is a reduced role for the state and an increased emphasis upon the use of market instruments). Policy on housing is a particularly good example of this shift. Although it has attracted little in the way of concerted attention from public policy analysts (other than to emphasise divisions within society and areas where a potential underclass is emerging) government policy has increasingly favoured the use of tax incentives in order to fashion urban regeneration and the construction of an expanded private rental sector. The difficulty in recognising such developments is due largely to the fact that (at least initially) the national level negotiations were seen (predominantly) as a social democratic approach to restructuring the economy. However, as they have developed (and this is by no means an inevitable path) they have assumed an increasingly conservative outlook (witness the budget decisions of 1998–99). At its most simplistic it is revealed in the very titles of the programmes: The Programme for National Economic Recovery, the Programme for Economic and Social Progress; the Programme for Competitiveness and Work and Partnership 2000. In a subtle, but nonetheless crucial fashion the agreements have shifted gradually (and ominously) toward a more conservative political position. It reinforces, once again, the importance of recognising that 'politics' is not just about what takes place in the Dáil. Policy is not simply the outcome of the choices of the respective political parties, but is shaped by global and national pressures emanating from the economy.

Presented in these terms, environmental policy can be understood as the outcome of a dynamic interaction between local, national and international developments. The crisis in regulation which manifested itself in the late 1980s condensed (and later institutionalised) these tensions in the formation of the EPA. Essentially, government recognised the need to accommodate the environmentalist critique of the 1980s without threatening the free market ethos which has become an increasingly prevalent feature of the politics of the Emerald Tiger. Put simply, these are tensions between the national and the international.

At the level of interaction between the national and the local, the principal theme of debate has been about political participation. For Janicke, or indeed Lafferty and Meadowcroft, it is here that national level negotiations (corporatist institutions) can offer environmental groups a potential political avenue through which participation may translate into change. I remain rather more circumspect, since this tends to 'isolate' the environmental dimension within political discussion. It may well be that the source of successful change lies within these institutional arrangements. However, as with the inclusion of voluntary organisations in Irish national level negotiations there is the problem that they may only pick from the 'crumbs' left from the 'main table'. The real source of change lies in constructing a new relationship between the economic and the environmental as it is interpreted within these institutions. In other words we crucially require a 'greening' of the core participants in the process, trade unions and business.

Conclusion

As we begin this millennium, it has become ever more clear that 'green issues' stubbornly refuse to relinquish their position of prominence on the Irish political landscape. Indeed, it seems that few days now pass without reference to the continuing problems of Ireland's ecology. From genetically modified foods to milk bottles, blue flags to interpretative centres, the letters' pages of the *Irish Times*, the radio and television are inundated with old and emerging environmental issues. Yet, it seems almost perverse to record that we are confronted by a confusing paradox; while environmental protest abounds, little in the way of a serious political debate has ensued. In part, this could be explained by the perception of the green issue as a matter of conservation or environmental protection: values which are largely peripheral to more pressing concerns of unemployment, crime or health. And if this issue does 'accidentally' appear on the political agenda, then it is an issue surely for the EPA and not the politicians. Indeed, one could be forgiven for believing that the agency functions as environmental equivalent to a tribunal; politicians, when confronted with thorny issue of green politics, find refuge in the line that 'this is the responsibility of the EPA'. To others, genuine concern about the state of Ireland's environment is all too frequently dismissed as the preoccupation of green fundamentalists, eco-warriors or people with 'too much time on their hands'. This implies that those who 'protest' or object do not operate in the real world, they do not realise that it is impossible to police the behaviour of all of the EPA's licensees or, that their demands for statutory, stringent, intrusive and punitive control are motivated simply by the desire to 'beat' the EPA and multi-nationals. This is a line of thought which ultimately espouses little more than a return to the pragmatic, incremental and loose regulatory style which summarily failed during the 1970s and 1980s.

While these views find support in vastly different political persuasions they coalesce on one familiar theme: that the environmental issue is not a political issue. But this is to ignore the extent to which environmental politics concerns issues which are crucial to the nature of Irish democracy: the appropriate role of state intervention; the ability of the Dáil to ensure that agencies such as the EPA are made accountable; anxiety about unemployment and the problems of social exclusion; sustainable development; green tourism; public concern about genetically modified foods; the fear of dioxin release; political protest at the location of landfills. These are real issues, environmental issues which are public issues and, consequently, they are *necessarily* political issues.

That Ireland's environmental regulatory framework has undergone substantial change in recent years is not in dispute. What is more, it would be misleading to suggest that important strides forward have not been achieved. Indeed, the EPA's willingness to provide access to environmental information and its determination to disseminate invaluable environmental data augur relatively well for the future. Moreover, the willingness on the part of planning authorities to insist on environmental impact statements and the subsequent improvements in the quality of those submissions also give grounds for optimism. These are initiatives which have also been complemented by the EPA's role in waste management, which has enabled the adoption of more stringent guidelines, improved monitoring and the inclusion of more robust procedures for post-closure care and maintenance. However, optimism needs to be tempered by caution; we should not be lulled into a false sense of security. Almost a decade has elapsed since a 'super agency' designed to drag Ireland's environmental regulation into the twenty first century was formed, and yet the optimism which accompanied the legislative passage of the EPA has largely dissipated.

By far the most important source of consternation is not the inadequacy of any particular element of the regulatory framework, but the persistence of policy style which has, more often than not, tended to accentuate the importance of making concessions and articulate the virtues of avoiding confrontation. It is an administrative style which remains incremental and ad hoc, where 'accommodations to reality' remain the order of the day. It is not the EPA's use of new technology, innovative policy instruments or new practices which have been derided here, since in many ways they have been innovative. The crux of the problem stems not so much from design but execution.

Those unfamiliar with the workings of the EPA, or the complexity of environmental regulation, have assumed that the formation of the agency signalled a radical departure in environmental regulation. It is a view which ignores the fact that, in many ways, there is less in the way of change than continuity with the regulatory patterns of the 1980s. The EPA's operating ethos is to 'persuade, cajole and encourage' the adoption of new eco-friendly approaches, rather than impose stringent environmental regulations. It is this 'throwback' to the regime of the 1980s, which has incurred the ire of the environmental lobby. The guiding thread between the government's rhetorical commitment to conserve the environment, its legislative promise to uphold environmental standards and the agency's commitment to monitor and enforce licence conditions, lies in the tendency to downplay the political influence of the planning, agricultural and industrial lobbies. These had, and continue to have, an important bearing upon how policy is formed, maintained and, ultimately, implemented.

It has been a prominent factor in the more significant deficiencies which persist in the regulatory framework: the contentious role of BATNEEC, an alarmingly low level of integration and the sub-division of responsibilities between the EPA and the planning authorities which has resulted in 'grey areas' emerging in planning and environmental control. The explanation of these difficulties has been sought in this book within the political dynamics which shaped the formation and implementation of the EPA. It has not been about the institutional characteristics of the EPA, or its personnel, but the politics which surround it.

From the outset, the EPA represented more than a simple realisation on the part of the Irish State that its environmental regulatory regime was inadequate. Indeed, as its institutional structure and operating rationale reveal, the motivation which lay behind the formation of the EPA was not one exclusively concerned with protecting the environment. Amid a crisis of confidence in environmental regulation toward the end of the 1980s the government recognised that, if it was to successfully revamp the legislative framework, and subsequently de-politicise the environmental policy domain, then it would need to construct a new environmental discourse which would sit comfortably within the free market rhetoric of the Emerald Tiger. The government's most pressing task was therefore to assuage the anxieties of the environmental lobby, diffuse potential areas of conflict and resistance and to remove from the public realm important questions about the relationship between economic growth and the environment.

There were two critical elements to this overall strategy. First, if the public's perception that industry was not being adequately controlled was to be dispelled, then the agency's independence would have to be at the forefront of the political debate. Second, the legislative powers of the EPA would have to reflect a significant shift from previous environmental regimes.

That reform was necessary was not in doubt. The challenge presented by community and environmental groups had served to question the influence of the planning, industrial and agricultural lobbies. Moreover, the extent of cross-party support for change indicated the pressing political need to replace a myriad of confusing legislative arrangements with a single agency to provide a co-ordinated, integrated and fully transparent approach to environmental regulation. Indeed, the very structure of the EPA, its obligations, operating rationale and regulatory philosophy, bear full witness to the scars of a struggle over the increasingly discredited environmental policy regime of the 1980s.

The EPA is not the outcome of a shift toward a new ecological modernity. Neither does it represent a radical overhaul of policy. Rather, it is an institutional manifestation of an attempt by the Irish State to accommodate the environmentalist critique of the 1980s without threatening the free market ethos which has underpinned the politics of the Emerald Tiger. A protracted upsurge in environmental protest had enveloped the Irish State in a struggle to construct a political discourse on the environment which was concerned not with embracing ideas about environmental degradation, justice, or democracy but with the complicated task of constructing a consensus around an *acceptable level of environmental pollution*. That this policy has singularly failed to effect a successful overhaul of environmental regulation in Ireland can be seen in the resurgence of environmental protests in recent years. For the protests of Merrell Dow or Raybestos Manhatten in the 1980s read genetically modified foods, architectural heritage or Mullaghmore in the 1990s. The participants in this unfolding drama may have changed, the stage upon which it has been enacted transformed, but it is the familiarity of the script which remains the more perturbing.

BIBLIOGRAPHY

Adshead, M, 1996. 'Beyond Clientelism: Agricultural Networks in Ireland and the EU', *West European Politics,* Vol. 19, No. 3, pp 583–608.

Allen, R., and T Jones, 1990. *Guests of the Nation; People of Ireland versus the Multi-Nationals,* Earthscan, London.

Baker, S., 1988. 'Dependent Industrialisation and Political Protest: Raybestos Manhatten in Ireland', *Government and Opposition*, Vol. 22 No. 3, pp 352–358.

Balchin, P. (ed.). 1996. *Housing Policy in Europe.* Routledge.

Barrett, A., and J. Lawlor, 1996. 'Solid Waste: Should We, Can We, Continue Relying on Landfill', *Irish Planning and Environmental Law Journal*, Vol. 3, No. 4, pp 73–79.

Barrett, A., Lawlor, J and S. Scott, 1997. *The Fiscal System and the Polluter Pays Principle: A Case Study of Ireland.* Ashgate.

Barry, U, and P. Jackson, 1989. 'Women's Employment and Multi-Nationals in Ireland: The Creation of a New Female Labour Force', in D. Eleson and R. Pearson (eds), *Women's Employment and Multi-Nationals in Europe*, Macmillan, London.

Beck, U, 1992. *Risk Society: Towards a New Modernity.* Sage.

1995. *Ecological Politics in the Age of Risk.* Polity.

1998. 'Politics of Risk Society' in Franklin, J, (ed), *The Politics of Risk Society.* pp 9–23.

Blackwell, J, and F. J. Convery, 1983. *Promise and Performance: Irish Environmental Policies Analysed.* Resource Environmental Policy Centre. Dublin.

Brassil, D, 1996. 'The Interface between Planning and IPC – After Masonite', *Irish Planning and Environmental Law Journal*, Vol. 3, No. 1, pp 20–23.

1996a. 'Waste Management and Land Use Planning – An Overview of the Issues in the Context of the Dublin Waste', *Irish Planning and Environmental Law Journal*, Vol. 3 No. 4, pp 53–55.

Buchanan, J. and G, Tullock. 1975. 'Polluters, Profits and Political Response: Direct Control versus Taxes', *American Economic Review*, 65, pp 139–47.

Byrne, D. 1996. 'Environmental Impact Statements: Their Role in the Process of Pig and Poultry Units in Ireland 1990–93', *Irish Planning and Environmental Law Journal*, Vol. 1, No. 3, pp 123–129.

Carson, R, 1965. *Silent Spring*. Harmondsworth, Penguin.

Christoff, P. 1996. 'Ecological Modernisation, Ecological Modernities', *Environmental Politics*, Vol. 5, No 3, pp 476–500.

Chubb, B., 1970. *The Government and Politics of Ireland*. Dublin, Gill and Macmillan.

Collins, K., and D. Earnshaw, 1992. 'The Implementation and Enforcement of European Community Environment Legislation', *Environmental Politics*, Vol. 1, No. 4, pp 213–249

Collins, T, 1991. 'Conflict Resolution as an Environmental Management Strategy' in Feehen, J, (ed,). *Environment and Development in Ireland*. The Environmental Institute, Dublin.

Coyle, C, 1994. 'Administrative Capacity and the Implementation of EU Environmental Policy in Ireland', in Baker, S, Milton, K and S. Yearley (eds) *Protecting the Periphery: Environmental Policy in the Peripheral Regions of the European Union*. Frank Cass, London.

Derham, J., 1995. 'IPC Licensing of Mining Activities', *Irish Planning and Environmental Law Journal*, Vol. 2, No. 4, pp 127–131.

1998. 'Integrated Pollution Control Licensed Industrial Activities – Waste Management and Groundwater', pp 1–14 EPA, Wexford.

1999. 'Providing for Environmental Liabilities in Integrated Pollution Control Licensed Operations', EPA. Wexford. pp 1–7.

Dobson, A., 1990. *Green Political Thought*. Unwin Hyman.

1996. 'Democratising Green Theory: Preconditions and Principles' in Doherty, B and M de Geus (eds)

Dodd, V.A., and W. S. T. Champ, 1983. 'Environmental Problems Associated with Intensive Animal Production Units, With Reference to the Catchment Area of Lough Sheelin', in Blackwell and Convery (eds)

Environment Council. 1980. *A Policy for the Environment*. Stationery Office, Dublin.

Doherty, B and M de Geus, 1996. *Democracy and Green Political Thought: Sustainability, Rights and Citizenship*. Routledge.

Doyle, A., 1998. 'IPC Licences', *Irish Planning and Environmental Law Journal*, Vol. 5, pp 152–155.

Doyle, B, 1996. 'Some Aspects of the Relationship of the Environmental Protection Agency with Local Authorities and An Bord Pleanála' *Irish Planning and Environmental Law Journal*, Vol. 3, No. 4, pp 155–58.

Dryzek, J, 1987. *Rational Ecology: Environment and Political Economy*. Basil Blackwell. Oxford.

1990. *Discursive Democracy: Politics Policy and Political Science*. Cambridge University Press.

Duffy, N., 'Fulfilling the Conditions of an Integrated Pollution Control licence', *Irish Planning and Environmental Law Journal*, Vol. 2, No. 3, pp 87–91.

Eckersly, R., 1992. *Environmentalism and Political Theory*. UCL Press.

1993. 'Disciplining the Market: Calling in the State: Four Competing Models for Integrating the Economy and the Environment'. ECPR workshop paper. 2–8 April. Leiden. Netherlands.

(ed,). 1995. *Markets, the State and the Environment: Towards Integration*. Macmillan.

Environmental Protection Agency. 1996. *Integrated Pollution Control Licensing: A Guide to Implementation and Enforcement in Ireland*. Wexford.

1997. *Report on IPC Licensing and Control*. Wexford.

1999. *Report on an Investigation of Recent Developments at Silvermines Tailings Management Facility, Co Tipperary*. pp 1–7.

Fanning, R, 1978. *The Irish Department of Finance, 1922–58*. Institute For Public Administration, Dublin.

Fenlon, R. M, 1983. 'The Water Pollution Control Act: An Evaluation', in Blackwell, J, and F. Convery (eds).

Flynn, T, 1996. 'Wind Farm Development – The Environmental and Planning Issues', *Irish Planning and Environmental Law Journal*, Vol. 3, No. 4, pp 143–151.

Friedman, M, 1962. *Capitalism and Freedom*. University of Chicago Press.

Fry, J, 1996. 'Assessing the Impact of Assessment', *Irish Planning and Environmental Law Journal*, Vol. 3, No. 4, pp 152–154.

1996b. 'The EPA's Draft Guidelines on EIA and Advice Notes' *Irish Planning and Environmental Law Journal*, Vol 3, No. 1, pp 11–12.

Galligan, Y, 1995. 'Editorial' *Irish Planning and Environmental Law Journal*. Vol. 2, no 1.

1996. 'Editorial: Masonite Inspector's Report' *Irish Planning and Environmental Law Journal*, Vol 3, No 1, 1996.

1996b, 'Editorial: An Bord Pleanála's Report, 1995' *Irish Planning and Environmental Law Journal*, Vol. 3, No. 3. p. 142.

1997. *Irish Planning Law and Procedure*. Roundhall Sweet & Maxwell.

Giddens, A, 1998. 'Risk Society: The Context of British Politics', in J. Franklin (ed,), *The Politics of Risk Society*. Polity Press, pp 23–35.

Government of Ireland. 1993. *Ireland: National Development Plan 1994–1999*, Stationery Office, Dublin.

Gobeyn M. J, 1993. 'Explaining the Decline of Macro-Political Bargaining Structures in Advanced Capitalist Societies', *Governance: An International Journal of Policy and Administration*, Vol. 6, pp 3–22.

Goodin, R, 1992. *Green Political Theory*. Cambridge. Polity.

Grabodsky, P, 1995. 'Governing at a Distance: Self-Regulating Green Markets', in Eckersly, R, (ed.),

Grist, B, 1997. 'Wildlife Legislation – the Rocky Road to Special Areas of Conservation Surveyed', *Irish Planning and Environmental Law Journal*, Vol. 4, No. 3, pp 87–95.

Grove-White, R, 1993. 'Environmentalism; A New Moral Discourse for Technological Society?' in Milton, K, (ed,).

1998. 'Risk Society, Politics and and BSE' in J. Franklin, (ed,), pp 50–54.

Hajer, M., 1995. *The Politics of Environmental Discourse: Ecological Modernisation and the Policy Process*. Clarendon Press. Oxford.

Harney, Mary, 1991. 'The Irish Environmental Protection Agency', in Feehen, J. (ed.), *Environment and Development in Ireland*. The Environmental Institute, Dublin.

Hayek, F, 1944. *The Road to Serfdom*. Routledge and Kegan Paul.

Helm, D, and D. Pearce., 1991. 'Economic Policy Towards the Environment: An Overview' in D. Helm (ed.,), *Economic Policy Towards the Environment*. Blackwell. Oxford.

Hickie, D, 1997. *Evaluation of Environmental Designations in Ireland*. The Heritage Council. Second edition.

Higgins, M. D, 1982 'The Limits of Clientilism: Towards an Assessment of Irish Politics' in C. Clapham (ed.), *Private Patronage and Public Power*. Frances Pinter. London.

Hildebrand, P. M, 1992. 'The European Community's Environmental Policy, 1957 to '1992': From Incidental Measures to an International Regime?' *Environmental Politics*, Vol. 1, No. 4, pp 13–43.

Inter-Departmental Environment Committee. 1977. *Report to the Minister for the Environment on Pollution Control*, Stationery Office, Dublin.

Institute for Industrial Research and Standards. 1977. *The Role of the IIRS in the Environmental Field*, IIRS, Dublin.

Jänicke , M., 1990. *State Failure: The Impotence of Politics in Industrial Society*. Polity Press. Cambridge.

1996. 'Democracy as a Condition for Environmental Policy Success: The Importance of non-Institutional Factors'in Lafferty and Meadowcroft (eds,).

Jordan, A, 1993. 'Integrated Pollution Control and the Evolving Structure of Environmental Regulation in the UK', *Environmental Politics*, Vol. 2, No. 3, pp 405–427.

Keohane, K, 1987. *Dependent Industrialisation, Crisis Transfer and Crisis Displacement: The Transnational Pharmaceutical Industry and the Irish Republic*, unpublished MA Thesis, UCC.

1989. 'Toxic Trade-Off: The Price Ireland Pays for Industrial Development', *The Ecologist*, Vol. 19, No. 4, pp 144–146.

King, D. S., 1987. *The New Right: Politics, Markets and Citizenship*. Macmillan.

Lafferty, W. M., and J. Meadowcroft., 1996. *Democracy and the Environment: Problems and Prospects*. Edward Elgar, Cheltenham.

Lash, S, Szerszynski, B and B. Wynne (eds), 1996. *Risk, Environment and Modernity*. Sage.

Lash, S, and J. Urry, 1987. *The End of Organised Capitalism*. Cambridge, Polity Press.

Lee, J J, 1990. *Ireland 1912 1985*, Cambridge University Press, Cambridge.

Leech, B C, 1989. *The Administrative and Technical Arrangements for Environmental Management in Ireland*. Vols. 1 and 2, Unpublished PhD Dissertation, Trinity College, Dublin.

Leonard, H J, 1988. *Pollution and the Struggle for the World Product; Multinational Corporations, Environment and International Comparative Advantage*. Cambridge University Press, Cambridge.

Leavy, T, 1997. 'The Rural Environmental Protection Scheme' *Farm and Food*, Spring.

Lowe, P and S Ward, 1998. *British Environmental Policy and Europe*. Routledge.

Industrial Development Authority. 1989. *IDA Industrial Plan 1978–1982*, IDA, Dublin.

MacLean, I., 1994. On IPC & Planning, BAT, Monitoring and Enforcement' *Irish Planning and Environmental Law Journal*, Vol. 1, No. 2, p. 82.

Maguire, B, 1995. 'Recent EU Developments', *Irish Planning and Environmental Law Journal*, Vol. 2, No. 3, pp 117–8

Malcomson, A P W, 1974. 'Absenteeism in Eighteenth Century Ireland', *Irish Journal of Social and Economic History*, pp 15–35.

Maguire, J, 1996. 'The Tears Inside the Stone: Reflections on the Ecology of Fear', in Lash, Szerynski and Wynne (eds,), pp 169–189.

Marsden, T, Murdoch, J, Lowe, P, Munton, R, and A Flynn, 1990. *Constructing the Countryside*. UCL Press. London.

Marsden, T, Lowe, P, and S Whatmore, 1990. *Rural Restructuring: Global Process and their Responses*. David Fulton Publishers, London.

Marsh, D, and R. A. W. Rhodes. 1992. *Policy Networks in British Government*. Oxford University Press.

McCarthy, E, and S Yearley, 1995. 'The Irish Environmental Protection Agency: The Early Years', *Environmental Politics*, Vol. 4, No. 4, pp 258–264.

McCormick, J, 1991, 'Environmental Politics', in Dunleavy, P. Gamble, A and G. Peele (ed). *Developments in British Politics 4*. Macmillan.

1991a. *British Politics and the Environment*. Earthscan. London.

McGrath, B, 1996. 'Environmentalism and Property Rights: The Mullaghmore Interpretative Centre Dispute', *Irish Journal of Sociology*, Vol. 6, pp 25–48.

McDonald, F. J, 1990. 'Can Europe Really Help?', in *Environmental Protection and the Impact of European Community Law*. Papers from the Joint Conference with the Incorporated Law Society of Ireland.

McLennan, G, 1992. *Marxism, Pluralism and Beyond*. Polity Press.

Milton, K, (ed,), 1993. *Environmentalism: The View From Anthropology*. Routledge.

Meehan, D., 1994. Integrated Pollution Control – New Licensing Arrangements', *Irish Planning and Environmental Law Journal*, Vol. 1, No. 2, pp 72–82.

1995. 'Planning Decisions and Integrated Pollution Control', *Irish Planning and Environmental Law Journal*, Vol. 2, No. 1, pp 30–34.

1996. 'The Waste Managment Act 1996: the Last Green Bottle', *Irish Planning and Environmental Law Journal*, Vol. 3, No. 2, pp 59–67.

Mullally, G, 1993. *Challenging the Illusion of Powerlessness; Social Movements, the Transnational Pharmaceutical Industry and the State in the Republic of Ireland*, Unpublished MA Thesis, University College Cork.

O'Donnell, R, 1998. 'Social Partnership in Ireland: Principles and Interpretations', in O'Donnell, R and J. Larragy (eds,), *Negotiated Economic and Social Governance and European Integration*. pp 84–105.

Offe, C, 1992. 'Bindings, Shackles, Brakes: On Self Limitation Strategies', in Honneth, A, McCarthy, T, Offe, C and A Wellmer (eds). *Cultural – Political Interventions in the Unfinished Project of Enlightenment*. MIT Press. London.

O'Riordan, T, 1987. *Risk Management and Hazardous Waste: Implementation and the Dialectics of Credibility*. International Institute for Applied Systems Analysis.

1996. 'Democracy and the Sustainability Transition' in Lafferty, W M and J. Meadowcroft, (eds,) pp 140–157.

O'Riordan, T, Kemp, R and M Purdue. 1988. *Sizewell B: An Anatomy of an Inquiry*. Macmillan.

O'Toole, F, 1994. *Black Hole, Green Card: The Disappearance of Ireland*. New Island Books. Dublin.

Paehlke, R, 1996. 'Democracy, the Environment and Public Opinion in Western Europe', in Lafferty, W. M and J. Meadowcroft, (eds), pp 18–39.

Peace, A, 1993. 'Environmental Protest, Bureaucratic Closure: The Politics of Discourse in Rural Ireland', in K. Milton, (ed.).

1994. Chemicals, Conflicts and the Irish Environmental Protection Agency, *Cork Environmental Alliance News*, No. 9, pp 17–22.

Pepper, D, *Modern Environmentalism*. Routledge.

Regan, S, 1994. 'Advice and Training in Agriculture to Meet the Change in the Direction of Farming', in M. Moloney (ed) *Agriculture and the Environment: Proceedings of a Conference on the Integration of EC Environmental Objectives with Agricultural Policy held in the RDS from March 9–11, 1994*. Royal Dublin Society, Dublin.

Richardson, J. J, and G. Jordan, 1982. *Policy Styles in Western Europe*. George Allen and Unwin.

Saward, M, 1996. 'Must Democrats be Environmentalists?' in Doherty, B and M de Geus, (eds).

Scannell, Y, 1982. *The Law and Practice Relating to Pollution Control in Ireland*. Graham and Trotman. London.

1990a. 'Impact of EC Water Pollution Directives in Ireland', in *Environmental Protection and the Impact of European Community Law*. Papers from the Joint Conference with the Incorporated Law Society of Ireland.

1990b. 'Legislation on Toxic Waste Disposal in Ireland', in *Environmental Protection and the Impact of European Community Law*.

Papers from the Joint Conference with the Incorporated Law Society of Ireland.

1993. 'Agriculture and Environmental Law' in Feehan, J, and F. Convery (eds).

1995. *Environmental and Planning Law in Ireland*. The Round Hall Press, Dublin.

Schaefer, G.F, 1991. 'The Subsidiarity Principle and European Environmental Policy' in *Subsidiarity: The Challenge of Change: Proceedings of the Jacques Delors Colloquium*, European Institute of Public Administration, Maastricht.

Schmitter, P C and G. Lembruch, 1979. *Trends in Corporatist Intermediation*. Beverly Hills Sage.

Sharp, R, 1998. 'Responding to Europeanisation: A Governmental Perspective' in Lowe and Ward (eds), pp 33–57.

Sherwood, M, 1994. 'The Role of the EPA in Agriculture', in M. Maloney (ed).

Simmons, G, 1997. 'Locus Standi to Challenge Planning Decisions', *Irish Planning and Environmental Law Journal*, Vol. 4, No. 3, pp 96–100.

Skou-Anderson, M, 1994. *Governance by Green Taxes: Making Pollution Prevention Pay*. Manchester University Press. Manchester.

Smith, A, (1776) (1828) *An Enquiry into the Nature and Causes of the Wealth of Nations*. Edinburgh, Adam and Charles Black.

Streeck, W, 1992. *Social Institutions and Economic Performance: Studies of Industrial Relations in Advanced Capitalist Economies* London, Sage.

Taylor, G., 1996. 'A Question of Interpretation: The Politics of an Environmental Dispute at Mullaghmore'. *Irish Journal of Sociology*, Vol. 6, pp 48–56.

1996a. 'Labour Market Rigidities, Institutional Impediments and Managerial Constraints: Some Reflections on the Recent Experience of Macro-Political Bargaining in Ireland', *Economic and Social Review*, Vol 27, No. 3, pp 253–277.

1998. 'Conserving the Emerald Tiger: The Politics of Environmental Regulation in Ireland', *Environmental Politics*, Winter, Vol. 7, No 4, pp 53–74.

1998a. 'Me Thinks Thou Dos't Protest too Much': Public Registers, Oral Hearings and Environmental Democracy in Ireland', *Irish Planning and Environmental Law Journal*, Vol. 5, No. 4, pp 143–152.

(ed,) 2001. *Issues in Irish Public Policy*. Irish Academic Press.

2001a. 'Public Policy and the Changing Archictecture of the Irish Political Landscape', in Taylor (ed,).

Urry, J, 1990. *The Tourist Gaze: Leisure and Travel in Contemporary Societies*. Sage. London.

Van der Straaten, J, 1994. 'A Sound European Environmental Policy: Challenges, Possibilities and Barriers', *Environmental Politics*. Vol. 1, No. 4, pp 65–81.

Ward, H, Samways., D and T. Benton, 1990. 'Environmental Politics and Policy', in Dunleavy, P, Gamble, A and G Peele (ed,). *Developments in British Politics 3*, Macmillan. London.

Weale, A, 1992. *The New Politics of Pollution*. Manchester University Press.

1996. 'Environmental Regulation and Administrative Reform in Britain' in G. Majone., (ed), *Regulating Europe*. Routledge.

Welsh, I, 1993. 'The Nimby Syndrome and its Significance in the History of the Nuclear Debate in Britain', *British Journal for the History of Science*, 26 (1) pp 15–32.

1995. *Nuclear Power: Generating Dissent*. London. Routledge.

Wynne, B, 1993. 'Methodology and Institutions: Value as Seen from the Risk Field', in J. Foster (ed), *Valuing Nature? Economics, Ethics and Environment*. Routledge. pp 135–155.

1996. 'May the Sheep Safely Graze: A Reflexive View of the Expert Lay Knowledge Divide', in Lash *et al*, pp 44–83.

Yearley, S., 1993. 'Standing in for Nature: The Practicalities of Environmental Organisations' Use of Science' in K Milton (ed).

Notes

1. The author would like to take the opportunity to thank R Edmondson, M Millar, F McLoughlin and A O'Reilly for comments on previous drafts.
2. The chapter title owes a great deal to the phrase used by J. Critchley and cited originally in Richardson and Jordan, 1982.
3. For more detail on Clientilism and Irish politics see M. D Higgins, 1982, on pluralism and its variants see G Mclennan, 1992 and on policy networks and policy communities see D. Marsh and R. A. W. Rhodes, 1992. M Adshead, 1996 situates the role of the Irish Farmers Association in these debates.
4. Richardson and Jordan have argued that it is possible to identify a 'characteristic' governmental policy style by establishing the manner in which the making of policy (and its implementation) is confirmed in two principal criteria: whether governments are active/reactive in their approach to problem solving and whether they are inclined to reach a consensus with organised groups (Richardson and Jordan, 1982, p. 13).
5. For an accessible introduction to the themes associated with the rise of environmentalism during this period see R Eckersly, 1992; D. Pepper, 1996; A Dobson, 1990.
6. The IIRS was set up, under the authority of the Department of Industry and Commerce, to facilitate industrial growth through technological research and development. It had no statutory function in the area of environmental policy but in the absence of clearly defined environmental standards the IIRS performed the function of advising government and local authorities on acceptable norms for industrial practise in relation to effluents, emissions and hazardous waste. The IDA commissioned the IIRS to report on the environmental implications of a number of projects which were subject to IDA grants. However, despite the fact that these reports were commissioned by one public body and carried out by another public body, the reports themselves were never made public (see Allen and Jones, 1990).
7. The phrase has its origins in the statements of Griffin, J in

[8] *Readymix v. Dublin County Council* where the finding was that in general the obligation was on the developer to describe the proposed development and the planning authority's function in assessing this information was that of 'watchdogs not bloodhounds'. Cited originally in Scannell, 1995, p. 180.

There has been little written about the Mullaghmore incident, one of Ireland's most protracted and acrimonious environmental disputes. McGrath (1996) has attempted to situate it within debates about the commodification of the countryside, suggesting that the landscape becomes just another product we purchase or consume; interpretative centres are part of a government policy which seeks to (re)define the landscape in terms of ability to generate revenue. It's an interesting line of thought, but in McGrath's case deeply flawed. See Taylor, 1996 for a reply. These are also themes touched upon by O'Toole who argues that landscape is no longer to be understood as a function of place or space, but to be aestheticised into a narrative story '... the grid of stories is designed to be plugged into on a journey around Ireland for tourists ... History is suspended in a commodified sense of place' (O'Toole, 1995, p. 40). Of more interest in this tradition is the work of J. Urry, 1990 and Marsden, T *et al*, 1990 & 1993. As the legal case dragged on (not a particularly uncommon feature of modern Ireland) various protest groups emerged to contest the local, political highground. The eye of the storm shifted from the preservation of the Burren's unique ecology, to one of jobs, tourism and the problems of rural Ireland. Here, objectors to the interpretative centre argued that once the project was completed, jobs would be only seasonal, generally low paid and predominantly part-time.

[9] In O'Toole's opinion, the phenomenon of the 'interpretative centre' is part of a 'post-modern hyper-reality' in a land of 'recreated "heritage" and forgotten history' (O'Toole, 1995, p. 37). What is peculiar about Ireland is that it has become 'a post-modern society without ever fully becoming a modern one' (O'Toole, 1995, p. 35).

[10] Skou-Andersen credits the German economist, Lutz Wiche

with the development of the idea of economic growth in environmental protection. In his work on *Der Oko-Plan: Durch Unweltschutz zum neuen Wirtshaftwunder* (The Eco Plan: From Environmental Protection to New Economic Wonder) [1984] and *Die okologischen Milliarden* (The Ecological Miracle) (1986). Wiche argued that stringent forms of environmental regulation need not impose an undue burden upon the economy but rather could stimulate the levels of growth experienced in Germany in the 1950s and 1960s. See Skou-Andersen, 1994. It was a theme taken up by the influential Culliton report (1992) which argued that Ireland should exploit the potential to present itself as a destination for 'green tourism'.

[11] Weale notes that Sweden was an exception to the general rule that all nations followed a multiple permitting approach. The Swedish have had a system of integrated permitting in place since 1969. The system takes into account the totality of polluting impacts and the aim is to reduce pollution to the lowest optimal level. Weale also notes however, that the Swedish political culture, characterised by consensus and an acceptance of state intervention, provided the ideal conditions for an integrated approach, conditions not easily replicated elsewhere (Weale, 1992b, pp 97–99).

[12] Under section 82(1) a new activity needs a licence from the prescribed day. An established activity from the specified day. The prescribed day is May 11th, 1994 and includes activities classified as 1, 2, 4, 5, 7, 8, 10 and 11. For classes 3, 12 and 13 it was April, 1995 and for classes 6 and 9 it was Sept, 1996. For more detail on this issue see A Doyle, 1998.

[13] In the case of the UK, the policy of HMIP on monitoring was that preference should be given to continuous on-line real time instrumental monitoring where available. Departures from this requirement, and the proposed frequency of any intermittent monitoring, would be justified by the operator on BATNEEC grounds. This information was to be placed on the public registers. However, the absence of this information has made it extremely difficult to assess the agency's performance in achieving this objective (ENDS, 1994).

[14] Mr Justice O'Sullivan was of the opinion that the 'standard'

required by these instruments could only be read in the light of the 1992 Act, the 1990 Directive and the 1994 regulations. The closest the 1992 Act comes to establishing a standard is in the phrase 'prevention of danger to health or damage to property or for the preservation of amenities' Section 111 (1). However, he noted that section 111(2) allowed for the assessment of 'possible risk to the environment' from a potential release and that if the 'possibility of risk was eliminated, then it was unlikely the Act would make provision for the assessment of risk and the study of such risk' (S Gillane, *Irish Times*, 14 Dec, 1998, p. 26).

[15] For more detail on this see S. Gillane, Law Report, *Irish Times*, 14 Dec, 1998, p. 26.

[16] Planning law is a vast and complicated subject area. This section attempts to condense some of the more important themes and where possible avoids technical jargon. Attempts to reproduce complex issues in this manner is often done at the expense of important nuances in law. If you are interested in these issues you should consult E. Galligan, 1997 and Y. Scannell, 1995, pp 288–312.

[17] On this matter the Environmental Research Unit's findings appear to be rather more optimistic than the conclusion drawn by D Byrne, 1996.

[18] For reasons of brevity these points are condensed. For a more nuanced summary of the finer points of this case and the respective findings see Galligan, 1997, pp 100–103.

[19] One of the more significant omissions in the formation of the EPA was the failure to develop any formal structure through which environmental groups could object to the granting of licences. Rather, there exists the *possibility* that the EPA will conduct an 'oral hearing' chaired by one of its representatives. The status of such hearings remains unclear, and it may be that the EPA will act as arbiters of appeals on their own decisions. For more detail see Taylor, 1998b. This is also discussed below in the issue of environmental democracy.

[20] The author would like to thank Sarah Reenan for her stimulating thoughts on this topic.

[21] From 1985–91, 20 SPAs were designated covering 6,959

hectares (80% state owned). Designation has increased as a result of infringement proceedings brought against the Irish Government by the EU. Evaluation is difficult because no SPA has been managed as an SPA. Although protection was initially limited all are now part of the NHAs.

[22] This has also occurred in the UK where guidance notes have generally eschewed specifying particular techniques so that operators have a degree of flexibility in defining BATNEEC. However, later notes have suggested that the onus is upon the operators to be aware of best available techniques and that companies should not cite the guidance notes as a reason for not adopting improved techniques when available (ENDS, 1994, p. 41).

[23] This latter point relates to the questionable practice of determining tolerability levels. There is a possibility that this conflicts with the Act's requirement to prevent, minimise or render harmless pollution releases. It is also possible that this may favour new projects in relatively unpolluted places (ENDS, 1994, p. 57).

[24] The EU Directive on EIA was influenced by the procedures adopted in the US (National Environmental Policy Act, 1969). However, in the US there is a more of a focus on 'alternatives', either in terms of changes within the project or a no project scenario. See Fry, 1996b.

[25] For those such as Mrs Thatcher, the post-war decline of the UK economy was precipitated by an inordinate expansion in the welfare state and a corresponding increase in the political access granted to trade unions. The argument runs that powerful trade unions distort the balance of resources in society (they secure better wages and terms of conditions for their members than the market would otherwise sustain) and prevent management from making the tough decisions (particularly in the public sector where industries can always be bailed out by government). These are familiar views often expressed in 'outbursts' against the role of the public sector in the Irish economy

INDEX

Allen, R, and T Jones, 26, 30
An Bord Pleanala, 5, 7, 61–63

Balchin, P, 107
Barrett, A, and J Lawlor, 100
Barry, U, and P Jackson, 28
Beck, U, 120–1
Brassil, D, 14, 65, 98
Buchanan, J, and G Tullock, 36, 144
Byrne, D, 112–3

Carson, R, 15
clientilism, 10, 11, 13
Collins, K, and D Earnshaw, 44
Control of Farmyard Pollution Scheme, 83–6
Coyle, C, 22
Culliton, J, 35

Derham, J, 102–6
Dobson, A, 127, 142
Dodd, V.A, and W. S. T Champ, 77
Doherty, B and M de Geus, 143
Doyle, A, 103
Dryzek, J, 37

Eckersly, R, 37, 144
environmental democracy, 142–153
Ecological modernisation, 35–6, 38–40, 119–25.
Environmental Impact Assessment, 62–3, 111–113
Environmental Impact Statements, 57, 62–3; 112–113
Environmental Protection Agency;
     and agriculture, 73–88
     and BATNEEC, 10, 47–55, 88–92, 139
     and environmental auditing, 95
     and forestry, 83–85
     and IPC licensing, 40, 42, 48–55, 73, 76, 88–92, 94, 98
     and local authorities, 9, 16, 24–7, 55–57, 81, 100
     and planning 101–103, 106–109
     and waste management, 75, 96–106

environmental lobby groups, 2, 14, 21, 33, 53, 76, 83, 126–7
environmental management, 4
environmental policy;
    and bureaucratic failure, 37
    and the European Union, 21–5, 42–5, 76, 78–80, 87, 97–8, 105, 142–3
    and regulatory capture, 37
    and science, 118–134, 140
environmental protest, 5, 33, 137–9
environmental welfare economics, 36–7, 144

Fry, J, 111–112

Galligan, E, 58–60, 64, 66, 109–110
Giddens, A, 121
Gobeyn M J, 148
Goodin, R, 142
Grist, B, 80–2
Grabodsky, R, 38–9
Grove-White, R, 119

Hajer, M, 35.
Hayek, F, 144
Helm D and C Pearce, 136
Hickie, D, 81, 91
Higgins, M. D, 11

Industrial Development Agency, 17, 27–30, 67–8

Jänicke, M, 37, 145

Keohane, K, 26

Lash, S., Szerszynski., B and B.Wynne, 120–22
Lash, S., and J. Urry, 148
Lee, J J, 149
Leech, B C, 25
Leonard, H J, 26–28
Leavy, T, 75–80
Lowe, P, and S. Ward, 21, 43–45

MacLean, I, 4, 91–3
Maguire, B, 46, 50
Malcomson, A. P. W, 72

Maguire, J, 132
Marsh, D, and R. A. W Rhodes, 15
McDonald, F. J, 21
Meehan, D, 90, 98

National Heritage Areas, 76, 80

O'Donnell, R, 148–9
Offe, C, 122–4
O'Riordan, T, 123, 129, 131–2, 140
O'Riordan, T, Kemp, R and M Purdue, 128–30, 134

Planning, 58–63, 112–14
    exempted development, 65–7
    and waste management, 63–5
pluralism, 11, 146
polyarchy, 11
policy community, 11, 14, 15, 30
policy style, 12–14
public registers, 89

Regan, S, 31
Richardson, J. J, and G. Jordan, 13
Rural Environmental Protection Scheme, 76–80, 82–3

Saward, M, 142–3
Scannell, Y, 9, 16–19, 22–24, 48, 56, 62–3
Schmitter, P C and G Lembruch, 146
Sharp, R, 45–6
Sherwood, M, 4, 73, 75
Simmons, G, 114–15, 135
Streeck, W, 148

Taylor, G, 5, 18, 20, 26–30, 52, 67–8, 77, 91–95, 136–8, 149–50

Van der Straaten, J, 45

Waste recycling, 100
Weale, A, 15, 21, 36, 47
Welsh, I, 122
Wynne, B, 121–2, 125–6

Yearley, S, 122, 139